Weight-of-Evidence for Forensic DNA Profiles

STATISTICS IN PRACTICE

Series Advisors

Human and Biological Sciences
Stephen Senn
CRP-Santé, Luxembourg

Earth and Environmental Sciences
Marian Scott
University of Glasgow, UK

Industry, Commerce and Finance
Wolfgang Jank
University of Maryland, USA

Founding Editor
Vic Barnett
Nottingham Trent University, UK

Statistics in Practice is an important international series of texts which provide detailed coverage of statistical concepts, methods and worked case studies in specific fields of investigation and study.

With sound motivation and many worked practical examples, the books show in down-to-earth terms how to select and use an appropriate range of statistical techniques in a particular practical field within each title's special topic area.

The books provide statistical support for professionals and research workers across a range of employment fields and research environments. Subject areas covered include medicine and pharmaceutics; industry, finance and commerce; public services; the earth and environmental sciences, and so on.

The books also provide support to students studying statistical courses applied to the above areas. The demand for graduates to be equipped for the work environment has led to such courses becoming increasingly prevalent at universities and colleges.

It is our aim to present judiciously chosen and well-written workbooks to meet everyday practical needs. Feedback of views from readers will be most valuable to monitor the success of this aim.

A complete list of titles in this series appears at the end of the volume.

Weight-of-Evidence for Forensic DNA Profiles

Second Edition

David J. Balding

*University of Melbourne, Australia
and University College London, UK*

And

Christopher D. Steele

University College London, UK

WILEY

Library of Congress Cataloging-in-Publication Data

Balding, D. J., author.
   Weight-of-evidence for forensic DNA profiles / David J. Balding and Christopher D. Steele. – Second edition.
       pages cm – (Statistics in practice)
   ISBN 978-1-118-81455-0 (cloth)
   1. Forensic genetics–Statistical methods. 2. DNA fingerprinting–Statistical methods. I. Steele, Christopher D., author. II. Title. III. Series: Statistics in practice.
   RA1057.5.A875 2015
   614′.1 – dc23
                                                                                    2015008388

A catalogue record for this book is available from the British Library.

ISBN: 9781118814550

Typeset in 10/12pt TimesLTStd by Laserwords Private Limited, Chennai, India

Printed and bound in Singapore by Markono Print Media Pte Ltd

1   2015

Contents

Sections marked † are at a higher level of specialization and/or are peripheral topics not required to follow the remainder of the book. They may be skipped at first reading.

Preface to the second edition

This new edition, appearing about 10 years after the first, retains most of the aims, structure and content of the original. Both Chapters 8, dealing with low-template DNA (LTDNA) profiling in general, and 9, introducing our software likeLTD for the evaluation of LTDNA profiles, are new. The material on DNA profiling technology has been updated, and there is some new material on DNA transfer and brief introductions to novel genomics-based and related forensic technologies (Chapter 4). The statistical material on the beta-binomial and multinomial-Dirichlet sampling formulas (Section 5.3) has been simplified, and Section 6.5 on mixture interpretation has been substantially rewritten. There are minor edits throughout, including many new references to literature appearing over the past decade.

We thank those who made helpful comments and suggestions on the first edition, in particular John Buckleton. We hope to have accommodated all those suggestions here. We welcome further comments and corrections and propose to maintain a website with comments from others or any errors/omissions that we note, www.wiley.com/go/balding/weight_of_evidence. We hope that the book will remain useful for many years to come: while the technology will continue to evolve, the principles of evidence evaluation remain essentially the same.

Preface to the second edition

Preface to the first edition

Thanks are due to Kathryn Firth for drawing the figure on page 11 and to Karen Ayres and Renuka Sornorajah for providing helpful comments on a draft of the book. Lianne Mayor contributed some of the material for Section 7.2.2. Discussions with many colleagues and friends over more than 10 years have contributed to the ideas in this book: many forensic scientists have helped me towards some understanding of the laboratory techniques, and I thank Peter Donnelly for stimulating comments on, suggestions about and criticisms of my statistical ideas during the formative years of my interest in the field.

I am grateful to John Buckleton and Colin Aitken for sending me pre-publication manuscripts of Buckleton et al. [2005] and Aitken and Taroni [2004], respectively; at the time of writing, the published versions have not appeared.

The statistical figures in this book were produced using R, a software package for statistical analysis and graphical display that has been developed by some of the world's leading statisticians. R is freely available for multiple platforms, with documentation, at www.r-project.org. Other diagrams have been created by the author using xfig, interactive drawing freeware that runs under the X Window System on most UNIX-compatible platforms, available from www.xfig.org.

1

Introduction

1.1 Weight-of-evidence theory

The introduction of DNA evidence around 1990 was a breakthrough for criminal justice, but it had something of a 'baptism of fire' in substantial controversy in the media and courts over the validity of the technology and the appropriate interpretation of the evidence. DNA profiling technology has advanced since then, and understanding by lawyers and forensic scientists of the appropriate methods for evaluating standard DNA profile evidence has also improved. However, the potential for crucial mistakes and misunderstandings remains. Although DNA evidence is typically very powerful, the circumstances under which it might not lead to satisfactory identification are not widely appreciated. Moreover, new problems have arisen with low-template DNA (LTDNA) profiles, which can be subject to stochastic events such as drop-in and drop-out.

The report of Caddy et al. [2008] was commissioned by the UK Government in response to the controversy over the 2007 acquittal of a defendant charged with the 1998 Omagh bombing in Northern Ireland. It found the underlying science to be 'sound' and LTDNA profiling to be 'fit for purpose', while admitting that there was lack of agreement 'on how LTDNA profiles are to be interpreted'. We find those phrases to be mutually incompatible. Fortunately, much progress has been made since 2008, but the international controversy surrounding the legal process arising from the murder of Meredith Kercher in Perugia, Italy, in 2009, in which LTDNA evidence played a central role, highlights the challenges that can arise. We aim in this book to present the fundamental concepts required for interpretation of DNA profiles, including LTDNA. We will initially focus on the general issues concerning the measurement of evidential weight, develop the weight-of-evidence theory based on

Weight-of-Evidence for Forensic DNA Profiles, Second Edition.
David J. Balding and Christopher D. Steele.
© 2015 John Wiley & Sons, Ltd. Published 2015 by John Wiley & Sons, Ltd.
Companion Website: www.wiley.com/go/balding/weight_of_evidence

likelihoods and discuss some alternative probability-based approaches. We will then apply the theory to forensic DNA profiling.

The primary goal of this book is to help equip a forensic scientist charged with presenting DNA evidence in court with guiding principles and technical knowledge for

- the preparation of statements that are fair, clear and helpful to courts, and

- responding to questioning by judges and lawyers.

The prototype application is identification of an unknown individual whose DNA profile was recovered from a crime scene, but we will also discuss profiles with multiple contributors, as well as paternity and other relatedness testing, and consider profiles that are subject to drop-out and other consequences of LTDNA and/or degraded DNA. We assume the setting of the United States, the United Kingdom and Commonwealth legal systems in which decisions on guilt or innocence in criminal cases are made by lay juries, but the general principles should apply to any legal system.

We will introduce and develop a weight-of-evidence theory based on two key tenets:

1. The central question in a criminal trial is whether or not the defendant is guilty.

2. Evidence is of value inasmuch as it alters the probability that the defendant is guilty.

Although these tenets may seem self-evident, it is surprising how often they are violated. Focussing on the right questions clarifies much of the confusion that has surrounded DNA evidence in the past.

It follows from our tenets that evidential weight can be measured by likelihoods and combined to assess the totality of the evidence using the appropriate version of Bayes' theorem. We will discuss how to use this theory in evaluating evidence and give principles for, and examples of, calculating likelihoods, including taking into account relevant population genetic factors.

No theory ever describes the real world perfectly, and forensic DNA profiling is a complex topic. The weight-of-evidence theory developed here cannot be applied in a naive, formulaic way to the practical situation faced by forensic scientists in court. Nevertheless, a firm grounding in the principles of the theory provides:

- the means to detect and thus avoid serious errors;

- a basis for assessing approximations and simplifications that might be used in court;

- a framework for deciding how to proceed when the case has unusual features;

- grounds for deciding what information a clear-thinking juror needs in order to understand the strength of DNA profile evidence.

Fortunately, we will see that the mathematical aspects of the theory are not too hard. Of course assessing some of the relevant probabilities – such as the probability that a sample handling error has occurred – can be difficult in practice, reflecting the real-world complexity of the problem. Further complications can arise, for example, in the case of mixed DNA samples (Section 6.5). However, the same simple rules and principles can give useful guidance in even the most complex settings.

There exist other theories of weight of evidence based on, for example, belief functions or fuzzy sets. The theory based on probability presented here is the most widely accepted, and its philosophical underpinnings are compelling [Bernardo and Smith, 2009, Good, 1991]. So whatever is actually said in court in connection with DNA evidence, it should not conflict with this theory.

There has been debate about the appropriateness in court of using numbers to measure weight of evidence. We only touch on this argument here (Sections 6.3.4 and 11.4.5). It is currently almost universal practice to accompany DNA evidence by numbers in an attempt to measure its weight (but see Section 11.4.6), and so we focus here on issues such as which numbers are most appropriate in court, and how they should be presented.

1.2   About the book

Chapters 2, 3 and 11 are not scientifically technical and, for the most part, are not specific to DNA evidence. We therefore hope that lawyers dealing with scientific evidence, and forensic scientists not principally concerned with DNA evidence, will also find at least these chapters to be useful. Courtroom lawyers ignorant of the weight-of-evidence theory described in Chapters 2 and 3 should be as rare as theatre critics ignorant of Shakespeare, yet in reality, we suspect that few are able to command its elegance, power and practical utility.

We first set out the weight-of-evidence theory informally, via a simplified model problem (Chapter 2) and then more formally using likelihoods (Chapter 3). In Chapter 4, we briefly survey DNA-based typing technologies, starting with an introduction to autosomal[1] short tandem repeat (STR) typing, emphasizing possibilities for typing error, then moving on to other DNA typing systems, digressing briefly to discuss fingerprint evidence and finishing with some newer evidence types: methylation, RNA and phenotype and ancestry prediction from DNA. Next, we survey some population genetics theory relevant to DNA profile evidence (Chapter 5). These two chapters prepare us for calculating likelihoods for DNA evidence, which is covered in Chapters 6 (identification) and 7 (relatedness). In Chapters 8 and 9, we extend identification inferences to LTDNA profiles and give a brief introduction to our freely available software for LTDNA profile evaluation, likeLTD. Chapter 10 discusses some alternative probability-based approaches to assessing evidential strength: none of these methods is recommended but each has its merits. In Chapter 11, we discuss some basic fallacies in the evaluation of DNA profile evidence and briefly review the opinions of some UK and US legal and scientific authorities.

---

[1] The nuclear chromosomes excluding X and Y.

1.3   DNA profiling technology

For the most part, we will assume that the DNA evidence is summarised for reporting purposes as the lengths of STR alleles at multiple autosomal loci (typically 10–25), reported as the number of tandem repeats of a DNA motif, usually 4 base pairs (bp) in length. The final result at four of the loci might be reported as

STR locus:	D18	D21	THO1	D8
Genotype:	14,16	28,31	9.3,9.3	10,13

in which each pair of numbers at a locus indicates the numbers of repeats on the individual's two alleles. Although whole repeats are the norm, partial repeats sometimes occur (Section 4.1.1); the profile represented here is homozygous for a THO1 allele that includes a partial repeat: the 9.3 indicates 9 full copies of the 4 bp DNA motif plus a 3 bp partial repeat.

Autosomal STRs now form the standard DNA typing technology in many countries. Both the 13-locus Combined DNA Index System (CODIS) and 15-locus European standard set (ESS) were derived by expanding earlier, smaller sets of STRs, and both have been superseded by larger sets. Most of Europe currently uses a 16-STR system developed from the ESS, which includes the amelogenin sex-identifying locus. The GlobalFiler® set from Life Technologies-Applied Biosystems offers 22 STRs, amelogenin and a Y-indel marker [Life Technologies, 2014], while the PowerPlex® Fusion System from Promega offers 23 STRs and amelogenin [Promega, 2014]. The different systems have many loci in common, but alleles at a locus can have different flanking regions (Section 4.1) in different systems. For more details of current STRs in use and developments that have allowed the expansion of STR sets, see Phillips et al. [2014].

The process of typing STR profiles is introduced in Section 4.1 but is not covered in great depth in this book. For further details, see Butler [2010]. Rudin and Inman [2001] give a general introduction to both technical and interpretation issues. Although we emphasise STR profiles, the principles emphasised in the following apply equally to any DNA profiling system. Interpreting profiles from the haploid parts of the human genome (the Y and mitochondrial chromosomes (mitochondrial DNA, mtDNA)) raises special difficulties. These systems are introduced in Sections 4.2 and 4.3, and interpretation issues specific to them are discussed briefly in Section 6.4. In Section 4.5, we briefly discuss profiles based on single-nucleotide polymorphism (SNP) markers.

1.4   What you need to know already

Chapters 2, 3, and 11 have essentially no technical prerequisites. To follow Chapter 5, you should know already what an STR profile is, have a rudimentary genetics vocabulary (locus, allele, etc.) and know the basic ideas of Mendelian inheritance. In statistics, you should be familiar at least with the theory of the error in a sample estimate of a population proportion (binomial distribution). The reader with experience of calculating with probabilities will be at an advantage in Chapters 6 and 7, but few technical tools are required from probability theory. In Sections 5.4.1 and 10.3, familiarity with statistical hypothesis testing are assumed, but these sections are labelled with a †, which means that they can be skipped without adverse impact on your understanding of the remainder of the book. The most important tool for computing probabilities is the sampling formula (5.6), which is expressed in a remarkably simple recursive form that can be used repeatedly to build up complex formulas. We give examples of its use, which requires only an ability to add and multiply, and with practice, anyone should be able to use it without difficulty.

We do not provide a general introduction to statistics (for an introduction in forensic settings, see Aitken and Taroni [2004]) and give only a brief introduction to population genetics (Section 5.1). We believe that many complications and much confusion have arisen unnecessarily in connection with DNA evidence because of a failure to grasp the basic principles of assessing evidential weight. If one focusses on the questions directly relevant to the forensic use of DNA profiles, the number of ideas and techniques needed from statistics and population genetics is small.

While the central ideas are not very difficult, inevitably, there are special cases with their unique complexities. In addition, new ideas always take some time to absorb. Given some effort, this book should equip you with the basic principles for tackling any problem of interpreting forensic DNA evidence. The details of complex scenarios will never be straightforward, and no book can replace the need for thought, care and judgement on the part of the forensic scientist. The goal of this book is to complement these with some technical information and bring them to bear on the appropriate questions with guiding principles for assessing weight of evidence.

1.5   Other resources

Part of the reason for writing the book is to synthesise and extend our previous contributions to the forensic science and related literature in a coherent manner. In particular, Chapter 3 is a development of Balding [2000], Section 7.1 extends the paternity section of Balding and Nichols [1995] and Section 10.1 is based on Balding [1999]. Perhaps the most important feature of the book is the introduction of the population genetics sampling formula (Section 5.3), and its systematic application to various identification and relatedness problems. This draws in part on Balding and Donnelly [1995a], Balding and Nichols [1995] and Balding [2003], but some of the development is new here. Chapter 8 draws on the development in Steele and Balding [2014a].

There are several other books that deal with the statistical interpretation of DNA and other evidence. Aitken and Taroni [2004] gave a thorough introduction to the statistical interpretation of scientific evidence in general, including DNA evidence among other evidence types. Robertson and Vignaux [1995] also dealt with a range of evidence types and emphasised interpretation issues from a lawyer's perspective, giving less attention to technical scientific aspects; for example, they did not discuss population genetics. Evett and Weir [1998] is perhaps closest to this book, but the treatment of population genetics issues by these authors is different from ours, as is their approach to introducing the relevant statistical issues. Butler [2014] gave an introduction to the interpretation issues raised by STR profile evidence, while Buckleton et al. [2005] offered a more extensive treatment (second edition expected in 2015).

As far as we can see, there is no major philosophical difference between us and these authors: we all embrace the use of likelihoods and Bayes' theorem to evaluate evidence. We emphasise different aspects according to our individual perspectives, experience and target audiences. Our book develops the weight-of-evidence theory in general and from an introductory level, and its approach to population genetics issues is unique, while remaining concisely focussed on DNA profile evidence, without extensive related material.

Charles Brenner's 'Forensic Mathematics' website dna-view.com provides much information and discussion; see also the encyclopaedia article [Brenner, 2006]. Weir [2007] offered a more extensive, one-chapter summary of many issues. The International Society for Forensic Genetics http://www.isfg.org/ is the principal professional organisation for forensic genetics, and it sponsors the leading journal *Forensic Science International: Genetics*. Biedermann et al. [2014] presented a recent collection of articles on issues related to the interpretation of DNA profile evidence. The International Conference on Forensic Inference and Statistics is held every 3 years and in 2014 was in Leiden, Netherlands; the next meeting is expected to be in 2017 in North America.

2

Crime on an island

For most of this book, we will be thinking about one or other of the following problems:

- We have two DNA profiles, one from a known individual and one from a crime scene. How do we measure the weight of this evidence in favour of the proposition that the known individual is the (or a) source of the crime-scene profile?

- We have DNA profiles from two or more known individuals. Some of the relationships among them may be known, but at least one putative relationship is in doubt. For example, we may have a known mother and child, and a possible father–child relationship is in question. How do we assess the weight of the DNA evidence for or against the proposed relationship(s)?

We will see that there are many subtleties and potential complicating factors in answering these questions: even formulating the right questions precisely is not easy. We need to proceed in steps, starting with simplified problems and progressively adding more realistic features. In fact, before we even start to think about identification or relatedness evidence, we will look at some warm-up examples to get the brain into gear and thinking about how to evaluate information using probabilities.

2.1 Warm-up examples

2.1.1 *People* v. *Collins* (California, 1968)

An apparently reliable witness testified that the crime was committed by two individuals with the characteristics given in the following table. A couple fitting the

Weight-of-Evidence for Forensic DNA Profiles, Second Edition.
David J. Balding and Christopher D. Steele.
© 2015 John Wiley & Sons, Ltd. Published 2015 by John Wiley & Sons, Ltd.
Companion Website: www.wiley.com/go/balding/weight_of_evidence

description was charged with the offence. An instructor in mathematics was called as expert witness and suggested the probabilities shown.

Trait	Probability
Yellow car	1/10
Man with moustache	1/4
Woman with pony tail	1/10
Woman with blond hair	1/3
Black man with beard	1/10
Inter-racial couple	1/1000

The mathematics instructor multiplied these probabilities together and obtained a very small number. The defendants were found guilty. The conviction was overturned by the California Supreme Court, but it made errors in its analysis.

Many criticisms can be raised about the prosecution evidence in *Collins*, and many pages of academic literature have been devoted to raising them. Much of this literature is flawed in crucial respects. We will not analyse this in any detail here, but the example raises fundamental issues such as

- What numbers should be presented to a court to help jurors evaluate evidential weight?

- How should those numbers be interpreted?

- What possible errors should jurors be warned against?

2.1.2   Disease testing: positive predictive value (PPV)

Suppose that, although showing no symptoms, you decide to take a diagnostic test for a particular latent disease. Let us assume the following facts:

- about 1 in 1000 asymptomatic individuals have the latent form of the disease;

- the false positive rate is 0.01 (this is the proportion of positive outcomes among those taking the test who do not have the disease);

- the false negative rate is 0.05 (this is the proportion of negative outcomes among those taking the test who do have the disease).

The test result comes back positive. How worried should you be? What is the probability that you have the disease?

The answer is that even after the positive result obtained from a reliable test, it remains unlikely that you have the disease. The reason is illustrated in Figure 2.1. For every 100 000 individuals tested, on average, 1094 will test positive, but of these, only 95 will be true positives. That is, just under 9% of positive test results correspond to true positives, and so there is still over 90% probability that you do not have the

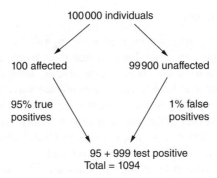

Figure 2.1 Schematic solution to the disease testing problem. The proportion of true positives among those who test positive is 95/1094, or about 9%.

disease. The test is not valueless: it has increased your disease probability from 1 in 1000 to about 87 in 1000, a big relative increase but, because of the small starting value, the final value is still not large.

Of course, this simple analysis does not apply exactly to you. In reality, there will be extra information, such as your age, sex, ethnicity, weight, lifestyle and general state of health, that affect your chance of having any particular disease. However, the analysis reveals an insight that is broadly valid in real situations: we must consider both the accuracy of the test *and* the prevalence of true positives in interpreting the test results.

The rare disease problem is closely connected with the weight-of-evidence problem: there are two possible 'states of nature', disease and no disease (compare with guilty and not guilty), and there is a diagnostic test that is reliable but can occasionally fail (cf. a DNA profile test that can occasionally result in a match by 'chance' or by error). The correct method of reasoning for rare diseases leads to results that at first are counter-intuitive for many people: it can do more harm than good to screen the general population for a rare disease, even when an accurate test is available.

The logic of the probability analysis is compelling, and its implications are now generally accepted for public health policy. The probability of having the disease given a positive test result is called the 'positive predictive value' (PPV) of the test.[1] It is a difficult quantity because it depends on the disease prevalence – and this will vary according to the factors listed above. In contrast, the false positive and false negative rates are easier to work with because they can be measured in the laboratory – so many focus on these even though they do not answer the relevant question for which the PPV is needed.

The analogous reasoning for DNA profile evidence also leads to some surprising conclusions, and after a period of struggle, they are becoming accepted in the courts. It is not the 'match probability' for the DNA profile test that ultimately matters, but the equivalent of the PPV, and this is a difficult quantity because it depends on the other evidence in the case.

---

[1] Also sometimes called the *precision* of the test.

2.1.3   Coloured taxis

Next, we consider an example of the sort of reasoning that led to the PPV for disease testing, but in the setting of a simplified evidence assessment problem. Suppose that 90% of the taxis in the town are green and the rest are blue. According to an eye-witness, the perpetrator of a 'hit-and-run' traffic offence was driving a blue taxi. We assume that the eyewitness testifies honestly but may have made a mistake about the colour of the taxi: it was dark at the time, and tests indicate that eyewitnesses mistake blue taxis for green, and vice versa, about 1 time in 10 under these conditions.

What is the probability that the taxi really was blue?

Both the false positive and false negative rates are 10%, but we have seen that these do not immediately provide the answer we seek. As a 'thought experiment', consider 100 such eyewitness reports. About 18 taxis will be reported to be blue, but only 9 of these are truly blue (Figure 2.2). So, after hearing an eyewitness report that the taxi was blue, the probability that it really was blue is 50%: the 'diagnostic' information that the taxi was blue exactly balances the background information that blue taxis are rare.

Had the eyewitness reported that the taxi was green, then the diagnostic information would support the background information to give an overall $81/82 \approx 99\%$ probability that the taxi really was green. This is a well-known example for evidence scholars and causes difficulty because it seems to imply that green-taxi drivers who offend will always get caught, yet blue-taxi drivers can drive dangerously with impunity. We do not discuss this conundrum further here, but note that it is not unreasonable to require more evidence to be persuaded that a rare event has occurred than for a common event.

2.2   Rare trait identification evidence

One of the methods scientists use to try to understand complex phenomena is to investigate simple models. Seeing what happens when simple models are tweaked can suggest insights into how the real world works. The models can be elaborated to

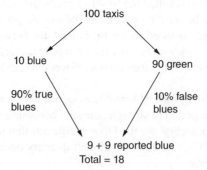

Figure 2.2   Schematic solution to the taxi problem. The probability that a taxi reported to be blue really was blue is 9/18 or 50%.

bring them a little closer to reality, but it is often the simplest models that give the most profound insights.

We will take this approach to assessing the weight of DNA evidence. An actual case involves many complications:

- how many previous suspects were investigated and excluded?

- what were the possibilities for a contamination error?

- are any of the suspect's close relatives possible culprits?

and many more. We cannot cope with all these complications at once. Instead, we will start with an imaginary crime on an island where life, and crime, is much simpler than in our world. Although unrealistic, analysis of the 'island problem' leads to profound insights. We will gradually add more features to bring the island closer to the real world. In so doing, we will learn new lessons about evidential weight.

Suspect = Culprit?

The island problem: facts summary

- All 101 islanders are initially equally under suspicion.
- The culprit has Υ.
- The suspect Q has Υ.
- The Υ-states of the other islanders are unknown.
- We expect on average about 1 person in 100 to have Υ.

What is the probability that the suspect is the culprit?

2.2.1   The 'island' problem

Consider a rare, latent trait: it could be a DNA profile but there is no need to be specific at this stage. Let us use the symbol Υ to denote the trait. A crime is committed on a remote island with a population of, say, 101. At first, there are no clues, and everyone

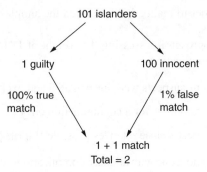

Figure 2.3   Schematic solution to the island problem. The probability that a matching individual is the culprit is 1/2 or 50%.

on the island is equally under suspicion. Then it is learned that the culprit possesses Υ, and a suspect Q is identified who has Υ. How convinced should we be that the suspect is guilty?

The answer to this question depends on, among other factors, how rare Υ is. Suppose that Q and the culprit (who may or may not be the same person) are the only people on the island whose Υ-status is known. A recent survey on the nearest continent, however, indicated that 1 person in 100 has Υ and we assume that islanders have Υ independently, with probability 0.01.

The facts of the island problem are summarised on page 11. A number of false 'solutions' have been presented in the evidence literature (for a discussion, see Eggleston [1983] and Balding and Donnelly [1995a]). The correct solution is sketched in Figure 2.3.

Even though Υ is rare, there is still only a 50% chance that Q is the culprit. In addition to the one individual known to match (Q), we expect one other individual on the island to match. In the absence of any further evidence, the probability that Q is guilty is 1/2. Just as for the case of the taxi reported to be blue, the background information that any particular individual is unlikely to be the culprit is exactly balanced here by the diagnostic information of the Υ-match.

2.2.2   A first lesson from the island problem

It is the probability that the defendant is guilty, given a match of Υ-states – the PPV – that is directly relevant to the juror's decision. The probability of guilt is difficult to work with in real crime cases because it can depend on many factors. In the simplified island setting, we are given that there is no other evidence and that all islanders are initially equally under suspicion, so we can ignore factors such as age, state of health and distance from the crime scene that complicate real-world crimes.

In general, if there are N people on the island other than the suspect Q and the probability that any one of them has Υ is p, then the formula for the probability that Q is guilty, given the Υ evidence, is

$$P(G|E) \equiv P(Q \text{ guilty given evidence}) = \frac{1}{1 + N \times p}. \qquad (2.1)$$

The notation '|' is mathematical shorthand for 'given', and we use P for 'probability', G for 'guilty' and E for 'evidence'. Thus $P(G \mid E)$ is a concise shorthand for 'the probability of guilt given the evidence'. Mathematical notation is remarkably ink efficient; that means it can seem opaque to those unfamiliar with it, but remember that it is just efficient shorthand, and it always stands for something that can be expressed (at much greater length) in words. We will see how to derive this equation in Section 3.2.

If $N = 100$ and $p = 1/100$, then

$$P(G \mid E) = \frac{1}{1 + 100 \times 1/100} = \frac{1}{2}.$$

The 1 in the denominator corresponds to the suspect, and the $100 \times 1/100$ corresponds to the (expected) one other Υ-bearer on the island.

Although it follows from (2.1) that the rarer is Υ, the higher is the probability that Q is guilty, the strength of the overall case against Q depends on *both* the rarity of Υ (i.e. on p) and the number of alternative possible culprits N – just as the probability of getting the disease following a positive test result depends on *both* the error rates and the prevalence of the disease.

Lesson 1 *The fact that Υ is rare (i.e. p is small) does not, taken alone, imply that Q is likely to be guilty.*

The unrealistic aspects of the island problem are immediately apparent. Nevertheless, Lesson 1 is 'robust': when we change the problem to make it more realistic, Lesson 1 still applies (see Section 3.4.1).

2.3   Making the island problem more realistic

Let us change some aspects of the island problem to investigate how different factors affect weight of evidence. To avoid getting too bogged down in complications, we will investigate different factors one at a time.

2.3.1   The effect of uncertainty about p

In practice, we will not know p and N exactly. Let us modify the island problem by attaching some uncertainty to p, measured by a standard deviation (SD) σ. The island problem formula (2.1) now becomes

$$P(G|E) = \frac{1}{1 + N \times (p + \sigma^2/p)}, \qquad (2.2)$$

which is always less than (2.1). We, therefore, have immediately the following important conclusion that, like Lesson 1, turns out to be robust.

Lesson 2 *Uncertainty about the value of p does not 'cancel out'. Ignoring this uncertainty is unfavourable to defendants.*

Numerical illustration: uncertainty about p

In the original island problem of Section 2.2.1, $N = 100$ and $p = 0.01$, and the probability of guilt is

$$P(G|E) = \frac{1}{1 + N \times p} = \frac{1}{1 + 100 \times 0.01} = 50\%.$$

If, now, we suppose that there is some uncertainty about p, say its SD is $\sigma = 0.005$, then the modified formula for the probability of guilt is

$$P(G|E) = \frac{1}{1 + N \times (p + \sigma^2/p)}$$

$$= \frac{1}{1 + 100 \times (0.01 + (0.005)^2/0.01)}$$

$$= \frac{1}{1 + 1 + 0.25} = \frac{4}{9} \approx 44\%.$$

So failing to acknowledge the uncertainty about p overstates the probability of guilt; here, 50% instead of the correct 44%.

Uncertainty about p can arise for a number of reasons. If knowledge about p comes from a survey, then there is uncertainty due to sampling: a different sample would have given a (slightly) different result. Moreover, there is uncertainty due to the possibility that the sample is unrepresentative (the islanders may differ from the population of the continent).

We will see in Chapter 6 that uncertainty about p is crucial to the correct interpretation of DNA evidence: uncertainty enters not only because of sampling error but also because DNA profile frequencies vary geographically and among ethnic/religious/social groups, and we never know exactly which is the correct reference group in a particular case or what are the profile frequencies in the relevant groups. The rarity of a DNA profile is assessed in part on the basis of data from 'convenience' samples rather than scientific random samples and in part on the basis of population genetics theory that holds at best only approximately in actual human populations. In practice, then, the σ^2/p in (2.2) is often much larger than the p, so that ignoring the former can be much more important than in the illustration above.

2.3.2   Uncertainty about N

The size N of the island population excluding Q may also be unknown, and uncertainty about its value also affects the probability that Q is guilty. However, in this case, the news is better: the effect of uncertainty about N is usually small, and ignoring it tends to favour Q. For example, if the island population size is expected to be N, but it could be either $N - 1$ or $N + 1$, each with probability ϵ, then the island problem formula (2.1) becomes

$$P(G \mid E) \approx \frac{1}{1 + Np(1 - 4\epsilon/N^3)} > \frac{1}{1 + Np}. \tag{2.3}$$

See Section 3.5.2 for further details.

Lesson 3 *Uncertainty about N, the number of alternative possible suspects, also does not 'cancel out', but the effect of ignoring this uncertainty is usually small and tends to favour defendants.*

2.3.3   The effect of possible typing errors

Now forget uncertainty, and assume again that we know p exactly. Suppose, however, that there is a probability ϵ_1 that an individual who does not have Υ will be wrongly recorded as having it (i.e. a false positive) and probability ϵ_2 for the other, false negative error. We assume that these probabilities apply for typing both Q and the sample from the culprit and that errors occur independently.

These assumptions are still unrealistically simple, but they allow a first insight into how the possibility of error affects evidential weight. The exact formula is a little complex, but an approximation appropriate here is given by

$$P(G|E) = \frac{1}{1 + N \times (p + \epsilon_1)^2/p}. \tag{2.4}$$

Notice that to a first approximation, ϵ_2 is irrelevant: since we have observed a match, the probability of an erroneous non-match is not important.

Numerical illustration: typing errors

If we suppose that there is a probability $\epsilon_1 = 0.005$ that any islander without Υ will be wrongly recorded as a Υ-bearer, then the probability of guilt is

$$\begin{aligned}
P(G \mid E) &\approx \frac{1}{1 + N \times (p + \epsilon_1)^2/p} \\[2mm]
&= \frac{1}{1 + 100 \times (0.01 + 0.005)^2/0.01} \\[2mm]
&= \frac{1}{1 + (1.5)^2} \\[2mm]
&= \frac{1}{3.25} \approx 31\%.
\end{aligned}$$

So ignoring the possibility of typing error leads to overstatement of the probability of guilt; here, 50% instead of the correct 31%.

We discuss the derivation of (2.4) in Section 3.5.4. This simple formula suffices for another important, robust, lesson.

> **Lesson 4** *The overall weight of evidence against Q involves adding together the probability of a true match and the probability of a false match due to a typing error.*

Laboratory and handling errors for DNA profile evidence are discussed further in Section 3.4.4. In general, the value of ϵ_1 is difficult to assess, usually more difficult than p since it depends strongly on the circumstances of a particular case. Ultimately, it is for the jury in criminal trials to assess the probability that some error has occurred, on the basis of the evidence presented to it. It is important that courts be given some idea of what errors are possible, how likely they are in the present case and what effect possible errors have on evidential weight.

2.3.4   The effect of searches

Forget, for the moment, uncertainty and errors, and focus on a new issue. In the island problem, the question of how Q came to the attention of the crime investigators was ignored. This is not as unrealistic as it may first appear: in practice, suspects are often identified on the basis of a combination of factors such as previous convictions, suspicious behaviour, criminal associates and so forth. Such reasons may not form part of the evidence presented in court, in which case, as far as a juror is concerned, Q 'just happened' to come to the attention of the authorities. Further, the legal maxim 'innocent until proven guilty' is usually interpreted to mean that before the evidence is presented in court, Q should be regarded as being just as likely to be guilty as other members of the population.

Suppose now that Q was identified on the basis of a search for Υ-bearers. That is, the islanders are examined in random order until the first one is found who has Υ. This individual is then accused of the crime. In addition to the facts listed in the summary on page 11, we now have the additional information that, say, k islanders have been investigated and found not to have Υ. In the original island problem, the reasons for first identifying Q were not based on Υ-possession.

> **Is $P(G|E)$, the probability that Q is guilty, higher or lower following a search, compared with the original island problem?**

There seem to be two reasons for believing that it should be lower:

1. the fact that Q was initially just one person in a random sequence of individuals searched means that he/she is less likely to be the culprit;

2. if you set out to find a suspect who has Υ, then the fact that Q has Υ is unsurprising and, therefore, of little or no evidential value.

It turns out that the probability that Q is guilty is *higher* following a search than in the original island problem. In fact,

$$P(G|E) = \frac{1}{1 + (N - k) \times p}, \qquad (2.5)$$

which is greater than (2.1). Many people, particularly scientists, find this result counter-intuitive, perhaps because of the two 'reasons' given above.

- The first 'reason' is easily dismissed: 'innocent until proven guilty' implies that every defendant should be treated, before the evidence is presented, as just another member of the population. This view is incorporated into the island problem by initially regarding every islander as equally under suspicion. So there is nothing special about a suspect identified on the basis of a search.

- To see that the second 'reason' is wrong, think about the case that Q was the last person searched: everyone else was inspected and found not to have Υ. Then the fact that Q has Υ is overwhelmingly strong evidence against him/her, as is reflected by the value $P(G|E) = 1$ that is obtained in (2.5) when $N = k$. The key is to keep attention fixed on the relevant question, which is not 'how likely is it that I will find a Υ-bearer if I look for one?' but 'given that a Υ-bearer has been found, how likely is it that he/she is the culprit?'.

The reason behind the correct formula is that each individual found not to have Υ is excluded from suspicion (remember that we are ignoring error here). The removal of this individual leaves a smaller pool of possible culprits, and hence, each remaining person in the pool becomes (slightly) more likely to be guilty. Notice that if the first person is found to have Υ, so that $k = 0$, then the original island problem formula is recovered. The fact that a search was intended then makes no difference to the strength of the evidence.

Lesson 5 *In the case of a search of possible culprits to find a match with crime scene evidence, the longer the search (i.e. the more individuals found not to match), the stronger the case against the one who is found to match.*

The important related issue of suspects identified through searches of DNA profile databases is discussed in Section 3.4.5.

2.3.5   The effect of other evidence

In the island problem, we assumed in effect that there was no evidence other than the Υ-evidence. In practice, of course, even if there is no further evidence that is directly incriminating, there will be background information presented to the jury, such as the location, time and nature of the crime that makes some individuals more plausible suspects than others.

> **Definition 1** *We write w_X for the weight of the non-Υ evidence against an individual X, relative to its weight against Q.*

In the original island problem, each w_X was equal to 1. A value $w_X < 1$ indicates that ignoring the Υ-evidence, X is less likely to be the culprit than is Q. As an example, suppose that other than the information about Υ, the evidence consists of the location of the crime and the locations of the homes of all the islanders. A juror may reason that individuals who live near to the crime scene are more likely to be the culprit than, say, individuals who live on the other side of the island. Such an assessment can be reflected by values of w_X greater than 1 for those who live nearer to the crime scene than Q and less than 1 for those who live further away.

When other evidence is taken into account, the island problem formula (2.1) becomes

$$P(G|E) = \frac{1}{1 + p \times \sum_{X=1}^{N} w_X}, \tag{2.6}$$

in which we introduce the mathematical symbol $\sum$ to denote summation. If all the w_X are equal to 1, then $\sum_{X=1}^{N} w_X = N$, and the formula reduces to the original island problem formula.

The role of the w_X in connection with assessing DNA evidence is discussed further in Section 3.4.2.

2.3.6   The effects of relatives and population subdivision

Even though we are initially ignorant about who on the island has Υ, the observation that Q has it can be informative about whether or not other individuals, such as relatives and associates, also have Υ. For example, if the population of the island is divided into 'easties' and 'westies', then the fact that Q, an eastie, has Υ may make it more likely that other easties also have Υ.

In the island problem, we assumed that Υ possession for different individuals was independent, so that one person's Υ-status carries no information about the Υ-states of other individuals. In practice, however, this 'learning' effect can be important, particularly for DNA profile evidence. Any particular DNA profile is very rare, but once that profile is observed, it becomes much more likely that other people, among the individual's relatives or ethnic group, also have it. It follows that what matters in practical cases involving DNA evidence is not p, the overall frequency of the profile, but the probabilities of the other possible culprits having the profile *given that Q* has it.

> **Definition 2** *We write m_X for the* match probability *for possible culprit X, which is the probability that X has Υ, given that Q has Υ.*

In the past, 'match probability' has been confused with the relative frequency of Υ in the population. This is incorrect because the concept of 'match' involves two individuals, not one. In the original island problem, we assumed independence of Υ-states so that m_X does equal p. For DNA profile evidence in real populations, however, relatedness and population subdivision mean that the match probability m_X exceeds the profile relative frequency p, often substantially. Thus confusing m_X with p is detrimental to defendants (see Chapter 6).

When these effects are taken into account, the island problem formula (2.1) becomes

$$P(G|E) = \frac{1}{1 + \sum_{X=1}^{N} m_X}. \tag{2.7}$$

If all the m_X are equal to p, then $\sum_{X=1}^{N} m_X = Np$, and (2.1) is recovered.

Lesson 6 *The strength of identification evidence based on an inherited trait depends (often strongly) on the degree of relatedness of the defendant with the other possible suspects.*

2.4   Weight-of-evidence exercises

Solutions start on page 185.

2.1   Suppose that weather forecasts are accurate: statistics show that rain had been forecast on about 85% of days that turned out to be rainy; whereas about 90% of dry days had been correctly forecast as dry. (To make matters simple, the weather has just two states, 'rainy' and 'dry'.) It follows that if rain is forecast, you may as well cancel the picnic, it is very likely to rain. Right or wrong? Assuming that the above figures are accurate, what is

(a) the probability of rain following a forecast of rain for Manchester, where rain falls, on average,[2] about 1 day in 5?

(b) the probability of rain following a forecast of rain for Alice Springs, where rain falls, on average, about 1 day in 100?

2.2   A diagnostic test for a latent medical condition has a false positive rate of 2 in 1000 and a false negative rate of 8 in 1000. Suppose that the prevalence of the condition in the general population is 1 in 10 000.

(a) What is the probability that someone who tests positive has the condition?

(b) Repeat (a), now assuming that the person tested comes from a high-risk group for which the prevalence is 1 in 250.

---

[2] These rainy day rates are just made up.

2.3 Suppose that the island has population 1001, all initially equally under suspicion. The suspect Q is tested and found to have Υ, whereas the Υ-states of the other 1000 inhabitants are unknown. Given that an islander has Υ, the probability that a particular unrelated islander also has it is $m = 5 \times 10^{-5}$, while m is 1 in 1000 for a cousin and 1 in 100 for a sibling. Ignore here the possibility of testing error and uncertainty about p.

(a) Suppose that Q is known to have no relatives on the island. What is the probability that he/she is guilty?

(b) Repeat (a), but now it is known that Q has one sibling and 20 cousins, and no other relatives, on the island.

(c) Repeat (b), but now the sibling of Q agrees to be profiled and is excluded.

(d) As a juror in the case, what assessment would you make about evidential strength if you were not given any information about the relatives of Q?

3

Assessing evidence using likelihoods

There are many other factors that could be introduced into the island problem in order to investigate their effect on evidential strength:

- what if the culprit is not, after all, the source of the DNA obtained from the crime scene?

- what about the fact that Q failed to produce a convincing alibi?

- what if the police accuse Q of every crime that occurs on the island?

- what if only a few individuals could have visited the crime scene during the time of the offence?

We will now introduce a general formula for quantitatively assessing evidence in the light of such factors. For a more advanced analysis, see, for example, Balding and Donnelly [1995a] and Dawid and Mortera [1996].

3.1 Likelihoods and their ratios

A likelihood is a probability of observed evidence given a hypothesis about that evidence. The absolute magnitude of the likelihood is not important, only its magnitude relative to the likelihoods for competing hypotheses. It is, therefore, common to express likelihoods as ratios relative to some baseline hypothesis. The match probability m_X, defined on page 18, is a special case of a likelihood ratio (LR). We first

Weight-of-Evidence for Forensic DNA Profiles, Second Edition.
David J. Balding and Christopher D. Steele.
© 2015 John Wiley & Sons, Ltd. Published 2015 by John Wiley & Sons, Ltd.
Companion Website: www.wiley.com/go/balding/weight_of_evidence

introduce LRs in general and later focus on LRs for DNA and other identification evidence.

Consider data D that is potentially informative about two rival (mutually exclusive) hypotheses, I and G (you can think of "innocent" and "guilty"). We define the LR $R_{I,G}(D)$ to be the ratio of the probabilities of the data under the hypotheses:

$$R_{I,G}(D) = \frac{P(D \mid I)}{P(D \mid G)}. \tag{3.1}$$

It does not matter whether it is I on top (numerator) and G below (denominator), or vice versa, as long as it is clear which has been used. Most practitioners use I as the reference hypothesis (denominator) so that large numbers represent strong evidence for the prosecution. There are advantages in the development of the theory (many formulas work out better) if G is the reference hypothesis, as in (3.1), in which case small values of the LR support the prosecution case. Throughout this book, we report LRs in the form of (3.1) and remind readers that our ratios are inverses of those discussed in much of the literature. Apologies for any inconvenience, but we tried doing it the other way and the formulas looked much worse.

In the disease testing example of Section 2.1.2, the data is the fact that the test result was positive. This has probability 0.95 if the hypothesis 'affected' (A) is true and probability 0.01 if 'unaffected' (U) holds. The LR is thus

$$R_{U,A}(\text{+ve test}) = \frac{P(\text{+ve test} \mid U)}{P(\text{+ve test} \mid A)} = \frac{1}{95}.$$

The positive test result supports hypothesis A much more than U. However, recall that because A may be a priori very unlikely, it does not necessarily follow that the +ve test suffices to make A more likely than U.

The value of $R_{I,G}(D)$ is a measure of the weight of evidence conveyed by D for I *relative to* its weight for G:

> $R_{I,G}(D) < 1$ means that D is less likely if I is true than if G is true. Whatever probabilities we previously assigned to the two hypotheses, observing D should lead us to decrease our belief that I is true relative to our belief in G.

If it is possible that neither I nor G is true, then D may be more likely under a different hypothesis. In that case, the probabilities for both I and G may decline as a result of observing D, but that for I will decline relatively further. However, if I and G are exhaustive (one of them must be true), then $R_{I,G}(D) < 1$ does imply that observing D decreases the probability of I and increases that of G.

An LR can be defined for any kind of data (or *evidence* or *information*). For example, Buckleton et al. [2005] noted that the absence of evidence (e.g. that a search of the defendant's home failed to find anything incriminating) is itself evidence for which an LR can in principle be calculated. In Section 6.1, there is a further discussion on LRs as measures of evidential weight.

3.2   The weight-of-evidence formula

Although we have given some intuitive explanation of (2.1) through (2.7), we have not yet explained how to derive such formulas.

Definition 3 *Let E_d stand for some evidence, while H_X and H_Q are the hypotheses that the perpetrator of the crime was, respectively, an individual X and the defendant Q. The LR comparing H_X with H_Q, given evidence E_d, is the ratio of how likely it is to have observed E_d under H_X to how likely E_d is under H_Q:*

$$R_X = \frac{P(E_d|H_X)}{P(E_d|H_Q)}. \tag{3.2}$$

Recall that for us, $R_X < 1$ when E_d supports H_Q more than H_X. With this definition, R_X can often be interpreted as a conditional probability (see Chapter 6).

Definition 4 *The* other evidence *ratio w_X, introduced informally on page 18, is the probability of H_X divided by the probability of H_Q, both evaluated in the light of E_o, all the evidence other than E_d. That is,*

$$w_X = \frac{P(H_X|E_o)}{P(H_Q|E_o)}.$$

Putting together the factors discussed in Sections 2.3.5 and 2.3.6, with the more general R_X replacing m_X, we obtain

---

The weight-of-evidence formula

$$P(G|E_d, E_o) = \frac{1}{1 + \sum_{X \in \mathcal{P}} w_X R_X}, \tag{3.3}$$

Formula (3.3) is a special case of a result in probability theory known as *Bayes' theorem*, in honour of the c18 clergyman Thomas Bayes. See any introductory probability textbook for more general formulations. The symbol $\in$ denotes "is a member of".

3.2.1   The population $\mathcal{P}$

The summation in (3.3) is over some population $\mathcal{P}$ of unprofiled individuals, assumed to include all the possible sources of the crime-scene profile (CSP) other than Q. Although $\mathcal{P}$ should include all realistic alternative suspects, there is some flexibility as to how many extra individuals are included. Often it might be appropriate to include in P all individuals aged, say, between 16 and 65 living within, say, 1 hour driving time of the crime scene. Alternatively, P might include all adult male residents of the nation where the crime occurred. However, P could include everyone on earth except Q, if desired: for a crime committed in Marrakesh, the value of w_X will be very close to zero when X is a resident of Pyongyang. This individual can be included in P, but the error resulting from simply ignoring all the residents of Pyongyang will usually be negligible.

3.2.2   Grouping the R_X

Although there is, in principle, a separate R_X for *every* person not excluded from being a possible culprit, in practice, there will be large groups of individuals for whom the evidence E_d bears the same weight and thus for whom R_X will take the same value. By grouping together members of P having approximately the same value of R_X, it will typically be satisfactory in practice to consider just a few distinct terms in the summation of (3.3).

For DNA evidence, there will typically be a small number of groups within which individuals bear approximately the same degree of relatedness to Q. For example, the population P of alternative culprits may be partitioned into

- identical (monozygote) twins of Q,
- siblings (including dizygote twins),
- parents and offspring of Q,
- relatives such as uncle, niece, grandparent and half-sibling of Q,
- cousins of Q,
- unrelated – same population, same subpopulation,
- unrelated – same population, different subpopulation,
- unrelated – different population.

The denominator of (3.3) can be rewritten with just one term for each of these groups (Section 3.5.3). If the value of R_X varies within a group, then applying the largest value to all members of the group avoids overstating the evidence against Q. This gives a simpler formula that provides only a lower bound on the probability of guilt, but the bound will often be adequate for practical use.

The notions of 'population' and 'subpopulation', although intuitive and widely used, are difficult to define precisely in practice, and hence, their usefulness may be disputed. These, and possibly other groups in the above list, can be combined leading to an even simpler formula that, if we continue to apply the largest R_X to all members of the combined group, gives a bound that is cruder but may still be satisfactory for use in court.

3.2.3   Application to the island problem

In the island problem setting, the evidence E_d can be summarised by

$$E_d = \text{'both } Q \text{ (suspect) and culprit are observed to have } \Upsilon\text{'.}$$

Since typing error is assumed impossible, E_d implies that suspect and culprit do both have Υ, in which case each R_X equals the match probability m_X. The LR is more general and, for example, can allow for the possibility that the evidence arose by handling error or fraud.

If Q is not the culprit, then we have observed two Υ-bearers on the island: the culprit and Q. Under the assumptions introduced on page 12, the probability that any two individuals have Υ is $p \times p = p^2$. On the other hand, if Q is the culprit, then we have observed only one Υ-bearer, and this observation has probability p. The LR for any possible culprit X is thus

$$R_X = \frac{p^2}{p} = p. \tag{3.4}$$

Substituting this value into (3.3), we recover (2.6), and in the case that all the w_X are equal to 1, we once again obtain the original island problem formula (2.1).

3.3   General application of the formula

3.3.1   Several items of evidence

When the evidence to be assessed, E_d, consists of two items, say, E_1 and E_2, the LR can be calculated in two equivalent ways, corresponding to the two possible orderings of E_1 and E_2, but these give the same result.

$$R_X(E_1, E_2 | E_o) = R_X(E_2 | E_o) R_X(E_1 | E_2, E_o) = R_X(E_1 | E_o) R_X(E_2 | E_1, E_o), \tag{3.5}$$

where, for example,

$$R_X(E_2 | E_1, E_o) = \frac{P(E_2 | H_X, E_1, E_o)}{P(E_2 | H_Q, E_1, E_o)}, \tag{3.6}$$

is the LR for evidence E_2 when E_1 is included with the background information, E_o. Further,

$$w_X(E_o) R_X(E_1 | E_o) = \frac{P(H_X | E_o)}{P(H_Q | E_o)} \frac{P(E_1 | H_X, E_o)}{P(E_1 | H_Q, E_o)}$$
$$= \frac{P(H_X | E_1, E_o)}{P(H_Q | E_1, E_o)}$$
$$= w_X(E_1, E_o).$$

Thus,

$$w_X(E_o)R_X(E_1, E_2|E_o) = w_X(E_1, E_o)R_X(E_2|E_1, E_o),$$

which means that applying the weight-of-evidence formula to E_1 and E_2 together, given background information E_o, gives the same result as applying it to E_2 when E_1 is included with E_o or to E_1 when E_2 is included with E_o. If you find some of the notation in this section difficult, do not worry: the take-home message is that the LR obeys common-sense rules of reasoning.

Apparently strong evidence may be of little value if it merely replicates previous evidence. If E_1 and E_2 are highly correlated (e.g. matches at tightly linked genetic loci or statements from two friends who witnessed the crime together and discussed it afterwards), then $R_X(E_1|E_2, E_o)$ and $R_X(E_2|E_1, E_o)$ may both be close to 1 (i.e. little evidential weight) even though both $R_X(E_1|E_o)$ and $R_X(E_2|E_o)$ indicate strong evidence. In this case, (3.5) implies that the joint weight of the two pieces of evidence is about the same as the weight of either piece of evidence taken alone. On the other hand, if the items of evidence are independent, given E_o, then

$$R_X(E_1, E_2|E_o) = R_X(E_1|E_o)R_X(E_2|E_o).$$

The LR for multiple items of evidence can similarly be built up by sequentially considering each item, in any order, but at each step, all previously considered items must be included with the background information. For example,

$$\begin{aligned} R_X(E_1, E_2, E_3|E_o) &= R_X(E_3|E_o)R_X(E_2|E_3, E_o)R_X(E_1|E_2, E_3, E_o) \\ &= R_X(E_3|E_o)R_X(E_1|E_3, E_o)R_X(E_2|E_1, E_3, E_o) \\ &= R_X(E_1|E_o)R_X(E_2|E_1, E_o)R_X(E_3|E_2, E_1, E_o) \\ &= \ldots \end{aligned}$$

The weight-of-evidence formula gives the same result if any one or two of E_1, E_2 and E_3 is included with E_o. For example,

$$w_X(E_o)R_X(E_1, E_2, E_3|E_o) = w_X(E_2, E_o)R_X(E_1, E_3|E_2, E_o).$$

It follows that the weight-of-evidence formula (3.3) can be used to assess all the evidence in a case. Suppose that the evidence can be allocated into four categories:

1. E_1 is information about the nature and location of the crime;

2. E_2 is an eyewitness description of the crime;

3. E_3 is the defendant's testimony;

4. E_d is the DNA evidence.

Initially, the jurors might assign values to the w_X based only on E_1. After hearing E_2, the juror can calculate $R_X(E_2|E_1)w_X(E_1)$. The probability of guilt $P(G|E_2, E_1)$ can be evaluated at this point, if desired. If E_3 is now taken into account, a new probability of guilt $P(G|E_3, E_2, E_1)$ can be calculated based on the values of $R_X(E_3|E_2, E_1)R_X(E_2|E_1)w_X(E_1)$ for all X. As noted above, these values can be regarded as $w_X(E_o)$ for the purposes of assessing the DNA evidence E_d, where E_o stands for all the non-DNA evidence, E_1, E_2 and E_3.

Numerical illustration: combining evidence

Suppose that two eyewitnesses give essentially the same testimony (E_1 and E_2) and that for either one of them, a juror assesses that the evidence is 10 times more likely if Q is the culprit than if X is guilty, so that

$$R_X(E_1 \mid E_o) = R_X(E_2 \mid E_o) = 0.1.$$

In one case, a juror may assess that the witnesses have discussed their evidence beforehand and smoothed over any differences, so that the second witness adds nothing to the first. Then $R_X(E_2 \mid E_1, E_o) = 1$ and

$$R_X(E_1, E_2 \mid E_o) = R_X(E_2 \mid E_1, E_o)R_X(E_1 \mid E_o) = 1 \times 0.1 = 0.1.$$

In another setting, the two witnesses may be regarded as independent:

$$R_X(E_1, E_2 \mid E_o) = R_X(E_2 \mid E_1, E_o)R_X(E_1 \mid E_o) = 0.1 \times 0.1 = 0.01.$$

Thus, the LR captures the intuition that two independent pieces of evidence are jointly more powerful (here $R = 0.01$) than two pieces of evidence that are essentially just replicates of the same information (here $R = 0.1$).

This logical analysis of evidence is particularly useful when some items of evidence are strongly incriminating while others are exculpatory. The key features are that the order in which different items of evidence is assessed does not affect the final answer, nor does how the evidence is categorised into items.

For most of this book, we will focus on applications of the formula to DNA evidence. We will assume that E_d refers to the DNA evidence and that E_o includes all other evidence, so that we regard the DNA evidence as being assessed last. This is for convenience and is not necessary; we will see in Section 11.4 that assessing the DNA evidence first has some advantages; in particular, it may then be reasonable to assume $w_X = 1$ for all X in $\mathcal{P}$.

Typically, most of the background information E_o, for example, information about alibis or eyewitness reports, has no effect on the LR $R_X(E_d|E_o)$. Note, however, that background information about the relatedness of X with Q, or their genetic ancestries, can be important in calculating LRs for DNA evidence.

Any individual can make their own assessment of the probability $P(G \mid E, B)$ that the defendant is guilty, based on the evidence E and any background information B that they feel appropriate. A juror's reasoning is, however, constrained by legal rules.

For example, although it may be reasonable to believe that the fact that a person is on trial makes it more likely that they are guilty, a juror is prohibited from reasoning in this way (because otherwise evidence may be double counted, see Section 11.1).

3.3.2   The role of the expert witness

The weight-of-evidence formula is very general and embodies the 'in principle' solution to the problem of interpreting DNA profile evidence, including the role of the non-DNA evidence and the effect of relatives, population variability and laboratory error. By 'in principle', we mean that it points out the quantities that need to be assessed and how they should be combined.

One important feature of (3.3) in connection with DNA evidence is that it helps demarcate the roles of jurors and expert witnesses. Ultimately, it is for jurors to assess evidential weight, but (3.3) indicates that a DNA expert can be most helpful to clear-thinking jurors by guiding them with reasonable values for the R_X. The w_X reflect jurors' assessments of the non-DNA evidence and will not usually be a matter for the (DNA expert) forensic scientist.

In the next section, we consider various consequences of the weight-of-evidence formula for assessing single-contributor DNA profile evidence. We continue to assume that the R_X are given; we defer computing LRs until Chapter 6, after some necessary population genetics has been introduced in Chapter 5. Multi-contributor (or 'mixed') DNA profiles are discussed in Section 6.5.

3.4   Consequences for DNA evidence

3.4.1   Many possible culprits

Because DNA evidence is widely, and correctly, perceived as being very strong, cases in which there is little evidence against the defendant other than the DNA evidence often arise. In such cases, there may be large numbers of individuals who, if not for that evidence, would be just as likely to be the culprit as the defendant (in other words, many individuals X for whom w_X is close to 1).

Even if all the LRs are very small, this may not suffice to imply a high probability for the defendant's guilt since the bottom line of the weight-of-evidence formula (3.3) involves a summation, and the total of many small quantities may not be small. A juror told only that 1 in 1 million persons has this profile may incorrectly conclude that this amounts to overwhelming proof of the defendant's guilt. This error can be extremely detrimental to defendants when there are many alternative possible culprits, or substantial exculpatory evidence.

A chance match

In 1999, a man from Swindon, UK, was found from the national DNA database to have a six-locus short tandem repeat (STR) profile that matched the profile from a crime scene in Bolton, over 300 km away. Despite the distance and the fact that the man was disabled, the reported match probability of 1 in 37 million was sufficiently convincing for the man to be arrested. A full 10-locus profile disclosed non-matches, and he was released.

The press made much of this event, describing it as undermining the credibility of DNA evidence. The UK *Daily Mail* reported a senior criminal review official as saying

> 'Everybody in the UK who has ever been convicted on six-[locus] profiling will want to apply to us to have their convictions reviewed.'

The match probability was frequently misrepresented as, for example, in a local newspaper:

> 'He was told that the chances of it being wrong were one in 37 million',

while *USA Today* said

> '... matched an innocent man to a burglary – a 1-in-37 million possibility that American experts call "mind-blowing"'.

The incident was unfortunate for the man concerned, and perhaps the police can be criticised for not taking the distance and disability into account before making the arrest. However, the incident has no adverse implications for DNA evidence. Indeed the match probability implies that we expect about two matches in the United Kingdom (population $\approx$60 million), and there could easily be three or four.

Because only a minority of individuals in the population currently has their DNA profile recorded, the extra matches would not be expected to be observed in any given case, but it is unsurprising that this eventually occurred. An analysis based on the weight-of-evidence formula (3.3) would not have generated a high probability that this man was the source of the crime-scene DNA.

The weight-of-evidence fallacy: examples

Many commentators seem to take the view that the fact that the profile is rare (i.e. p is small) alone establishes guilt. Some examples of statements that seem to be based on this fallacy are

- 'There is absolutely no need to come in with figures like "1 in a billion", "1 in 10,000" is just as good'.

- 'The range may span one or two orders of magnitude, but such a range will have little practical impact on LRs as large as several million'.

- 'population frequencies ... 10^{-5} or 10^{-7}. The distinction is irrelevant for courtroom use'.

These statements are misleading because in the presence of many possible culprits, or strong exculpatory evidence, very small LRs may be consistent with acquittal, and differences of one or two orders of magnitude may be crucial.

3.4.2 Incorporating the non-DNA evidence

Consider two assault cases for which the non-DNA evidence, E_o, differ dramatically:

Two assault cases

Case 1 – Victim recognises alleged assailant and reports his name, Q, to police;

   – Q is found to have injuries consistent with the victim's allegation and cannot give a convincing alibi for his whereabouts at the time of the alleged offence;

   – Q is profiled and found to match the crime profile.

Case 2 – Victim does not see assailant and can give no useful information about him;

   – The crime profile is compared with DNA profiles from many other individuals until a matching individual Q is found.

   – Q lives in another part of the country has a good alibi for the time of the crime and no additional evidence can be found linking him to the alleged offence.

The overall cases against the defendant, Q, differ dramatically: in the first case, the evidence against Q seems overwhelming, whereas in the second, a jury would have to make careful judgements about the validity of the alibi, the possibility of travelling such a distance and the strength of the DNA evidence. In the weight-of-evidence formula, the difference between these two cases is encapsulated in different values for the w_X. A plausible allegation by the victim in Case 1 may lead a juror to assign small values of w_X to each alternative possible culprit X. Lacking such an allegation, and faced with strong alibi evidence, jurors in Case 2 may assign values > 1 to many of the w_X.

A juror may be reluctant to assign precise values to the w_X but can make broad distinctions between the moderately large and extremely small values that may be appropriate in these two examples. Of course, jurors may not use any numerical evaluation of the evidence, but if forensic scientists have a quantitative understanding of the key issues, they can be more helpful in formulating their presentation of evidence to the jurors.

As noted in Section 3.2, the w_X can be calculated using (3.3) sequentially, considering all the items of non-DNA evidence one at a time.

3.4.3   Relatives

Because DNA profiles are inherited, relatives are more likely to share a DNA profile than are unrelated individuals. Many commentators have taken the view that close relatives of the defendant need not be considered unless there is specific evidence to cast suspicion on them.

The weight-of-evidence formula shows this view to be mistaken. Consider the case outlined in the box (next page). It may be helpful to profile brothers in such cases, if possible. The brother may, however, be missing or refuse to co-operate.[1] It may not even be known whether or not the defendant has any brothers.

Although still more simple than realistic cases, the example serves to illustrate the general point that consideration of unexcluded close relatives may be enough to raise reasonable doubt about the defendant's guilt *even when there is no direct evidence to cast suspicion on the relatives.*

3.4.4   Laboratory and handling errors

If crime scene and defendant profiles originate from the same individual, the observation of matching profiles is not surprising. Non-matching profiles could, nevertheless, have arisen through an error in the laboratory or at the crime scene, such as an incorrect sample label or laboratory record, a contaminated sample, a software error in a computer-driven laboratory procedure or evidence tampering. The common practice of ignoring this possibility favours the defendant, although the effect is typically small.

---

[1] In an early DNA evidence case in the United Kingdom, the police arrested a brother of the principal suspect in order to compel him to give a DNA sample and hence exclude him from consideration as an alternative culprit. However, the arrest was ruled illegal by the court and the resulting DNA profile ruled inadmissible.

Relatives calculation: an example

There is direct DNA profile evidence against a defendant Q, but the DNA profiles of the other possible culprits – a brother of Q named B and 100 unrelated men – are not available. The non-DNA evidence does not distinguish between these 102 individuals, so that the 'other evidence' ratios w_X are all equal to 1. We will defer consideration of methods for calculating LRs until Section 6.2. Here, we will take the following values: for the brother, $R_B = 1/100$; for all other possible culprits, $R_X = 1/1\,000\,000$. Then

$$P(G|E) = \frac{1}{1 + 1/100 + 100/1\,000\,000} \approx 99\%. \tag{3.7}$$

The probability of the defendant's innocence in this case is about 1%. A juror may or may not choose to convict on the basis of this calculation: the pertinent point is that ignoring the brother would give a misleading view of evidential strength, leading to probability of innocence of only 0.01%. It is easy to think of similar situations, involving additional unexcluded brothers or other close relatives, in which the probability of innocence is substantial, even after apparently strong DNA evidence has been taken into account. For an actual case in which the possibility that a brother was the culprit had an important effect on the outcome of an appeal, see Section 11.4.4.

On the other hand, when the defendant is not guilty, ignoring the possibility of error is always detrimental to the defendant, sometimes substantially so. The observed match could have arisen in two ways:

(a) suspect and culprit happen to have matching DNA profiles and no typing error occurred;

(b) suspect and culprit have distinct DNA profiles, and the observation of matching profiles is due to an error in one or both recorded profiles.

Both (a) and (b) are typically unlikely. In many cases, (b) may be important, but (a) may be the focus of more attention, in part because error probabilities are difficult to assess. Even if error rates from external, blind trials are available, there will usually be specific details of the case at hand that differ from the circumstances under which the trials were conducted and that make it more or less likely that an error has occurred.

We saw in Section 2.3.3 the role of error probabilities under simple assumptions. Some broad conclusions of these analyses are as follows:

- In order to achieve a satisfactory conviction based primarily on DNA evidence, the prosecution needs to persuade the jury that the relevant error probabilities are small.

- If the probability of error (b) is much greater than the probability of matching profiles (a), then the LR corresponding to (a) is effectively irrelevant to evidential weight.

- What matters are not the probabilities of *any* profiling or handling errors but only the probabilities of errors that could have led to the observed DNA profile match.

Ignoring the possibility of false-match error is always detrimental to Q, sometimes substantially so. Thompson et al. [2003] argued that jurors may mistakenly believe that a small error-match probability can be ignored, misunderstanding the point that it is not the absolute but the *relative* magnitude of the false-match to the chance-match probabilities that determines whether or not the former can be safely neglected.

Some critics of the use of DNA profiling have ignored the final point. Dawid and Mortera [1998] pointed out that this error is similar to that highlighted in the 18th century by Price in response to an argument of Hume. Even if a printing error is more likely than winning the lottery, a newspaper report that your number has won the lottery should not be dismissed: it is not the probability of *any* printing error that is pertinent but only an error that led to your number being reported, and this is typically less likely than an accurate report of your number. In contrast, with typical DNA evidence, a chance match is so extremely unlikely that the alternative possibility of the profile of Q was observed to match the CSP through error or fraud may well be much more likely, though still unlikely in absolute terms.

3.4.5 Database searches

The United Kingdom has a national database of the DNA profiles of named individuals for criminal intelligence purposes. At the start of 2014, nearly 5 million people had their profiles recorded in the UK NDNAD (4 million men, 1 million women; 78% white, 7% black, 5% South Asian, 8% unknown), while the number of profiles from unsolved crimes was 0.4 million. In the 9 months, April-December 2013, there were 18 195 suspect-to-crime-scene DNA matches, including 132 murders and 368 rapes.[2]

The question thus arises as to the appropriate method for assessing the DNA profile evidence when the defendant was identified following a search through this database. The number of individuals involved in such a search, and even the fact that there was a search, may not be reported to the court. This is because intelligence databases consist primarily of the DNA profiles of previous offenders, and admitting that such a search has been conducted is thus tantamount to admitting previous convictions. It is important to know whether or not omitting this information tends to favour the prosecution.

---

[2] NDNAD annual report 2012–2013, available at https://www.gov.uk/government/publications/national-dna-database-annual-report-2012-to-2013.

In Section 2.3.4, we considered the related problem of a sequential search in the population of possible offenders in the setting of the island problem. As we discussed in Section 2.3.4, it is widely – but wrongly – believed that the fact that a DNA profile match is more likely when it results from a search means that the evidence is weakened by the search. Many commentators, including the US National Research Council (NRC, see Section 11.5) and the German Stain Commission (Schneider et al. 2009), have also made errors in connection with intelligence databases.

In applying the weight-of-evidence formula, details of the search such as the number of non-matches can be treated as part of the background information. We find that Lesson 5 still holds, and DNA evidence is usually slightly stronger in the database search setting than when no search has occurred. Thus, omitting information about the search tends to favour the defendant, although usually the effect is small. The intuition behind the database search result is twofold:

(a) the other individuals in the database were found not to match and, hence, are effectively excluded from suspicion, reducing the number of possible culprits;

(b) the observation of many non-matches strengthens the belief that the profile is rare.

As an illustration of (a), consider an enormous database that records the DNA profiles of everyone on earth. If the defendant's profile were the only one in this database to match the CSP, then the evidence against him would clearly be overwhelming.

Although the DNA evidence may be slightly stronger in the context of a database search, the overall case against the defendant may tend to be weaker because there may often be little non-DNA evidence. For details of the analysis and discussion, see Balding and Donnelly [1996] and Balding [2002].

In addition to searches based on complete profile matches, it is possible to search a database using either a partial profile or the profile of a close relative of the target individual (Section 7.3). In that case, the result of a database search is not simply a small number of matching profiles but a (possibly long) list of profiles ranked according to the quality of the match with the queried profile. Gill [2014a] argued that naive investigators who accuse the first-ranked individual in such a search and do not seek further evidence may generate false allegations. If investigators proceed so irrationally, then there will indeed be false allegations, but when those cases come to court, if the courts proceed rationally, the defendants will not be convicted. The fact that there was a search need play no role in their deliberations. Indeed, as we have discussed, the court need not know that there was a database search. In every case involving the identity of a source of crime-scene DNA, the question is whether the jurors are convinced by the prosecution claim that Q is the source, to the exclusion of all possible X. If the match probability is extremely small, as is typical for a good-quality single-source sample, then jurors should need little if any additional evidence to be convinced, but if the LR is not so extreme because it comes from a partial match or a match with a relative, they should require additional evidence. How Q came to be a suspected contributor is irrelevant to that question.

3.5   Derivation of the weight-of-evidence formula †

So far we have stated many results without derivation. In this section, we fill in some of the missing details, which should be of interest to those who have mastered a first course in probability.

3.5.1   Bayes' theorem

Suppose that we have evidence E and the two hypotheses confronting a juror in a criminal trial, which we have been denoting G (Q is guilty) and I ($= $ not G). For simplicity, we continue to compare hypotheses G and I, but see Section 6.1 for discussion on more appropriate hypothesis pairs for DNA evidence, and in Section 6.5, we discuss hypotheses dealing with multiple unknown contributors to a DNA sample.

Bayes' theorem describes how to update *prior* probabilities of G and I to take into account the information conveyed by E. Since exactly one of G and I is true, we must have $P(G) + P(I) = 1$, and Bayes' theorem is

$$P(G|E) = \frac{P(E|G)P(G)}{P(E|G)P(G) + P(E|I)P(I)}. \tag{3.8}$$

All the probabilities in (3.8) are conditional on background information E_o.

Although valid, (3.8) is not immediately useful for many types of evidence, including DNA, because $P(E|I)$ can depend on who is guilty, if not Q. It is, therefore, convenient to partition I into many possible events H_X, where X denotes an individual other than Q. Then $P(I) = \sum_X P(H_X)$ and

$$P(E|I)P(I) = \sum_X P(E|H_X)P(H_X).$$

Substitution in (3.8), and for consistency writing H_Q in place of G, leads to the weight-of-evidence formula (3.3).

3.5.2   Uncertainty about p and N

Equation (2.2), which is the island problem formula modified to take into account uncertainty about p, can be derived by taking expectations in both the numerator and the denominator of the LR (3.2). Let the proportion of Υ-bearers in the island population now be $\tilde{p}$, a random variable with mean (i.e. average) p and SD σ. The numerator of R_X is the probability that a particular islander is a Υ-bearer. If the value of $\tilde{p}$ were known, this would be the required probability. Since $\tilde{p}$ is unknown, we must use its average, or expected value, $E[\tilde{p}] = p$. Thus, the denominator of R_X is unaffected by the uncertainty. The numerator of R_X is the probability that two distinct individuals, Q and X, are Υ-bearers, which is the expectation of $\tilde{p}^2$, written as $E[\tilde{p}^2]$. From the definition of variance

$$\text{Var}[\tilde{p}] = \sigma^2 = E[\tilde{p}^2] - E[\tilde{p}]^2,$$

it follows that
$$E[\tilde{p}^2] = p^2 + \sigma^2,$$

and so (3.2) becomes
$$R_X = \frac{p^2 + \sigma^2}{p} = p + \frac{\sigma^2}{p}.$$

As we noted in Section 2.3.1, the magnitude of the effect of uncertainty about p on the value of R_X can be large: often σ^2/p is much larger than p. This is because of population genetic effects to be discussed in Chapter 5. Intuitively, the observation of a particular DNA profile makes it much more likely, because of common ancestry, that another copy of the same profile will also exist.

Whereas uncertainty about p affects the LRs R_X, uncertainty about N affects the prior probability, $P(G)$. Let the number of innocent islanders be $\tilde{N}$, a random variable with mean N. Conditional on the value of $\tilde{N}$, the prior probability of guilt is $P(G \mid \tilde{N}) = 1/(1 + \tilde{N})$, but since $\tilde{N}$ is unknown, we need to use the expectation:

$$P(G) = E[G \mid \tilde{N}] = E\left[\frac{1}{1 + \tilde{N}}\right].$$

Because $1/(1 + \tilde{N})$ is not a symmetric function of $\tilde{N}$ but is convex (the function curve lies below the straight line between any two points on it), it follows from a fundamental result of probability theory (Jensen's inequality) that whatever the probability distribution for $\tilde{N}$, provided that $E[\tilde{N}] = N$, the prior probability of guilt is never less than in the N known case. That is,

$$P(G) = E\left[\frac{1}{1 + \tilde{N}}\right] \geq \frac{1}{1 + N}.$$

The bound is tight unless the variance of $\tilde{N}$ is large. Returning to the example of Section 2.3.2, if

$$\tilde{N} = \begin{cases} N - 1 & \text{with probability} & \epsilon \\ N & \text{with probability} & 1 - 2\epsilon \\ N + 1 & \text{with probability} & \epsilon, \end{cases}$$

then

$$\begin{aligned} E\left[\frac{1}{1 + \tilde{N}}\right] &= \frac{\epsilon}{N} + \frac{1 - 2\epsilon}{1 + N} + \frac{\epsilon}{2 + N} \\ &= \frac{1}{1 + N} + \frac{2\epsilon}{N(1 + N)(2 + N)} \qquad (3.9) \\ &\geq \frac{1}{1 + N}. \end{aligned}$$

Using (3.9) for the prior probability of G in the weight-of-evidence formula leads to the approximation (2.3) for the island problem formula.

Because uncertainty about N does not affect the LRs, the higher prior probability of guilt implies a higher posterior probability than in the N-known case. Ignoring this uncertainty thus tends to favour defendants, though the effect is usually small. In practice, uncertainty about the population size is usually dealt with by replacing N with an over-estimate, which also tends to favour defendants.

3.5.3   Grouping the alternative possible culprits

We noted in Section 3.2 that for DNA identification evidence, it usually suffices to group together the possible culprits that have the same, or similar, degree of relatedness to Q and, hence, that have the same, or similar, value of R_X. If, for example, P is partitioned into three groups, a, b and c, and the smallest LR for the individuals in each group is R_a, R_b and R_c, respectively, then (3.3) becomes

$$P(G \mid E_d, E_o) \geq \frac{1}{1 + R_a \sum_{i \in a} w_X + R_b \sum_{i \in b} w_X + R_c \sum_{i \in c} w_X}. \qquad (3.10)$$

If the LRs are nearly the same in each group, then the bound in (3.10) will be tight. If also the w_X are the same over all i in each group, then we can write

$$P(G \mid E_d, E_o) \geq \frac{1}{1 + N_a R_a w_a + N_b R_b w_b + N_c R_c w_c}, \qquad (3.11)$$

where N_a, N_b and N_c are the numbers of individuals in each group. If these are unknown, then the comments concerning uncertainty about N in Section 3.5.2 apply to each of them.

3.5.4   Typing errors

Consider again the modification to the island problem discussed in Section 2.3.3, in which typing errors occur independently with probabilities ϵ_1 and ϵ_2. Here, if suspect and culprit are not the same person, then the evidence must have arisen in one of three ways:

- Both suspect and culprit have Υ, and no typing error occurred; this has probability $p^2 (1 - \epsilon_2)^2$.

- One of the two has Υ, the other does not but a false positive error occurred; this has probability $2p(1 - p)\epsilon_1(1 - \epsilon_2)$.

- Neither suspect nor culprit have Υ, and both were incorrectly typed; this has probability $(1 - p)^2 \epsilon_1^2$.

If suspect and culprit are the same person, then there are two ways to have observed the evidence:

- The suspect/culprit has Υ and was correctly typed twice; this has probability $p(1 - \epsilon_2)^2$.

- The suspect/culprit does not have Υ and was incorrectly typed twice; this has probability $(1 - p)\epsilon_1^2$.

Combining all these probabilities, we obtain

$$R_X = \frac{p^2(1-\epsilon_2)^2 + 2p\epsilon_1(1-p)(1-\epsilon_2) + (1-p)^2\epsilon_1^2}{p(1-\epsilon_2)^2 + (1-p)\epsilon_1^2} \approx \frac{p + 2p\epsilon_1 + \epsilon_1^2}{p}.$$

The approximation holds if p, ϵ_1 and ϵ_2 are all small.

3.6   Further weight-of-evidence exercises

Solutions start on page 182.

3.1 Suppose that the evidence E_d is that glass fragments were found on Q's clothing, and the refractive index matches that of a window broken during a crime. Assume here that the person who broke the glass is the perpetrator of the crime. Recall that according to our weight-of-evidence theory, we need to assess the LR, R_X, that is, the probability of observing E_d if some individual X committed the offence, divided by this probability if Q did it (assume that the effect of the other evidence E_o can be neglected).

   (a) Think of a specific individual X. What does a rational juror need to assess in order to calculate R_X?

   (b) How might the assessment described in (a) vary for different X?

   (c) What information might an expert witness reasonably be able to provide to assist the juror make these assessments?

3.2 This exercise is loosely based on the UK case R v $Watters$, briefly outlined in Section 11.4.4. We assume here and in the next question that the source of the crime-scene DNA profile is the perpetrator of the crime. Suppose that there are 100 000 unrelated possible culprits in addition to the defendant Q, and for each of these, we have $w_X = 1$ and $R_X = 1/87$ million. Assume that for each of the two brothers of Q, we have $w_X = 1$ and $R_X = 1/267$, and that there are no other close relatives of Q among the possible culprits.

   (a) What is the probability that Q is the culprit?

   (b) How does the probability in (a) change if we now exclude the two brothers ($w_X = 0$ for each of them)?

3.3 This exercise is loosely based on R v $Adams$ (Section 11.4.2). The prosecution case rested on a DNA profile match linking the defendant with the crime, for which a match probability of 1 in 200 million was reported. The defence case was supported by an alibi witness who seems not to have been discredited at trial. The victim stated in court that the defendant Q did not resemble the man who attacked her, nor did he fit the description that she gave at the time of the offence.

(a) Consider only the DNA evidence, and assume that the 1 in 200 million figure is appropriate. It was reported in court that there were about 150 000 males between the ages of 15 and 60 within 15 km of the crime scene. On the basis of this information only, what might be a reasonable value for the probability that Q is guilty? (Assume that these 150 000 males are all unrelated to the defendant.)

(b) Update this probability taking the alibi evidence into account, supposing that this evidence is four times as likely if Q is innocent than if he is guilty (irrespective of the true culprit).

(c) Now take the victim's evidence into account, supposing that this evidence is ten times as likely if Q is innocent than if he is guilty (whoever is the culprit).

(d) Now suppose that the figure of 1 in 200 million is overstated and that the correct value is 1 in 20 million. What approximate probability of guilt might you arrive at if you were a juror?

(e) It is easy to criticise the analysis of (a) through (d) as relying on unreasonable assumptions. What practical suggestions can you make to improve the analysis?

(f) How might you convey the implications of the above analysis to jurors; alternatively, explain a different approach that you would take to convey the weight of the DNA evidence to jurors.

3.4 Return to the original island (Section 2.2.1) with 101 inhabitants each of whom has (independently) probability $p = 0.01$ of Y-possession. If all 100 remaining islanders are untested, we calculated that the probability that Q is guilty is 50%.

Now we are told that Q was identified because he was found to be the only Y-bearer among the 21 islanders whose Y-status is recorded in the island's database. Assume that inclusion in the database does not in itself affect the probability of guilt for this crime.

(a) What now is the probability of guilt?

(b) Write down a general expression for the probability of guilt when the database size is $n + 1$. What values does your expression take when $n = 0$, $n = 20$ and $n = 100$?

3.5 Leaving the island again, consider a more realistic database search scenario in which most of the individuals in the database are not plausible alternative culprits for the crime, either because they were in jail at the time or because they live far from the crime scene.

(a) Describe, in general terms, the effect on the evidential weight of a match found via a search in such a database.

(b) How, using our weight-of-evidence approach, would we assess the situation in which someone thought very unlikely to be a possible culprit turned out to have a matching DNA profile?

(c) If the matching individual in (b) was definitively excluded from suspicion because he was in jail at the time of the offence, what alternative lines of enquiry are suggested?

4

Profiling technologies

So far in this book, we have been considering general principles for the evaluation of evidence, particularly rare-trait identification evidence. In this chapter, we begin the task of bringing these principles to bear on the DNA profile evidence in current forensic use. A full account of DNA profiling technology is beyond the scope of the book (see Section 1.3 for further references). However, error is always possible in any human endeavour, and the assessment of the possibility of error is an essential part of the juror's task and one that the forensic scientist should seek to assist as far as possible. Therefore, we need to consider the possible sources of error in different DNA profiling technologies and how these might affect evidence interpretation. The most common source of errors in connection with DNA evidence are routine handling and clerical errors. These are likely to be equally prevalent whatever the typing technology used and may often be high. A European collaboration exercise to attempt to coordinate mitochondrial DNA (mtDNA) typing [Parson et al., 2004] recorded that for over 10% of the samples, at least one error occurred in a participating laboratory. Of these, 75% were clerical errors or sample mix-up which, fortunately, are unlikely to lead to a false match.

4.1   STR typing

The basic steps in the production of a short tandem repeat (STR) profile from a biological sample are

1. extraction of DNA;

2. amplification via PCR (polymerase chain reaction);

3. separation of PCR products (alleles) according to sequence length;

4. detection via fluorescent dyes.

Weight-of-Evidence for Forensic DNA Profiles, Second Edition.
David J. Balding and Christopher D. Steele.
© 2015 John Wiley & Sons, Ltd. Published 2015 by John Wiley & Sons, Ltd.
Companion Website: www.wiley.com/go/balding/weight_of_evidence

Step 1 will vary according to the source material, which may include hairs; biological fluids such as blood, semen or saliva; or cellular material deposited on, for example, clothing, mobile phones, tools or cigarette lighters.

Step 3 is achieved by the separation of PCR products through electrophoresis; DNA is naturally negatively charged, so under an electric field, it travels in the direction of a positive charge. Larger DNA fragments are impeded more by the structure of the gel, so that the fragment size can be estimated (usually to the nearest base pair (bp)) from the time taken to travel through the capillary. After subtracting the lengths of flanking sequences, the number of tandem repeats (including partial repeats) can be inferred.

The amount of DNA that is electrokinetically injected into the capillary depends on a number of factors, some of which can be altered by the forensic practitioner: strength of electric field applied, injection time, DNA concentration in the sample and the ratio of the ionic strength of the sample and buffer [Rose and Jorgenson, 1988]. Many methods to enhance typing of low-template samples aim to alter one or more of these factors, to increase the amount of DNA that will be injected (see Chapter 8).

To achieve Step 4, the PCR primers have fluorescent dyes incorporated into their 5′ ends [Giusti and Adriano, 1993]. At the detector window of the capillary, an argon-ion laser is shone on the capillary, which causes the fluorescent dyes to emit visible light that is detected using an optical camera. The development of multiple dyes that fluoresce at different wavelengths, and charged-coupled device cameras that can detect fluorescence at multiple wavelengths simultaneously, has allowed the development of 'multiplex' STR kits that are now standard. One dye is used to detect a standard set of known DNA fragments added to each sample, called the allele ladder. This is useful to aid the conversion from the time of detection to fragment length. Each other dye can be used to detect multiple loci with non-overlapping allele size ranges. Commercial STR kits can now use five dyes to multiplex over 20 loci; for example, Promega's Fusion kit [Promega, 2014] includes 23 STRs plus amelogenin. Recently, a sixth (purple) dye has been developed, after changing the red dye wavelength [Phillips et al., 2014], which should allow for a further increase in the number of markers.

The fluorescence of each dye is measured in relative fluorescence units (RFU), recorded over time in a graphical output called an electropherogram (epg). Time is converted into fragment lengths in bp, usually using software provided with the typing machine. The signal from each dye is recorded on a separate panel of the epg, with alleles represented by peaks.

Figure 4.1 shows an epg together with allelic designations resulting from STR profiling of a buccal swab from a male individual. Each of the four panels of the epg shows the signal intensity over time for a different colour dye, converted into fragment length in bp indicated above each panel. Figure 4.1a ('Blue') records the individual's genotypes at four autosomal loci. Figure 4.1b ('Green') records the amelogenin sex-distinguishing locus (leftmost; here recording an XY genotype for a male) and three autosomal loci. Figure 4.1c ('Black') records a further four autosomal loci, while Figure 4.1d ('Red') records five more autosomal loci. Another panel ("Orange", not shown here) records the allele ladder, which is represented by vertical stripes on

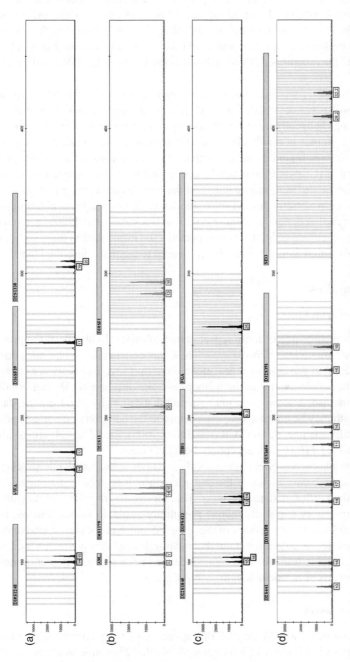

Figure 4.1   An electropherogram showing the STR profile of a single, male individual. Each panel corresponds to a different dye colour (from top to bottom): Blue, Green, Black and Red. The DNA sample was amplified with the NGM SElect^TM STR kit with 30 PCR cycles. The amplified fragments were separated on the ABI 3130xL Genetic Analyzer and analysed using the GeneMapper® 3.2 software. Image supplied courtesy of Cellmark Forensics.

the panels that are displayed here. The two peaks labelled 14 and 15 at the left end of the Figure 4.1a are separated by 4 bp, as expected for consecutive alleles at a tetranucleotide STR locus. Size resolution to the nearest bp is usually feasible under good conditions. Note that the rightmost peaks on the epg, corresponding to the longest alleles, tend to be lower than other peaks, and these are most susceptible to allelic 'drop-out' when the quantity of DNA analysed is small and/or the DNA is degraded (Chapter 8).

In Chapters 6 and 7, we will, for the most part, assume that the STR profiles have been recorded without error. For good-quality samples, such as that underlying Figure 4.1, this is usually a reasonable assumption: it is apparent from the figure that STR allele calling is normally clear-cut. However, there are a number of anomalies that may arise even for good-quality samples. More importantly, for low-template crime scene profiles (CSPs) that may also be contaminated and/or degraded, the above assumptions may be far from reasonable, and we will discuss appropriate analyses in Chapter 8. Here, we briefly discuss some anomalies that may arise in obtaining an STR profile. For further details, see Butler [2005] or Buckleton et al. [2005] and references therein. Sequencing of STR loci is expected to be introduced in the near future (see Section 4.6) which would replace the epg-based technology described here.

4.1.1   Anomalies

4.1.1.1   Microvariants

STR alleles that include one or more repeat units that differ slightly from the prototype repeat motif are called *microvariants*. These can consist of a single nucleotide insertion or deletion in one or more repeat elements or may involve a sequence change that does not alter allele length. In the former case, the different alleles can usually be reliably distinguished. However, a microvariant that differs by 1 bp from the standard repeat length may be confused with *adenylation*: the addition of a non-template nucleotide, usually adenine, during PCR (see Butler, 2005). Partial repeats are usually rare, the 9.3 allele at THO1 is an exception, being common.

Because electrophoresis in effect measures allele length, same-length microvariants are not distinguished: two alleles of the same length will be recorded as matching alleles even if they differ at sequence level. This means that recorded STR profiles have less discriminatory capability than is potentially available via sequencing, but this causes no problem for the validity of the recorded match and its associated match probability. In particular, a defence claim that the profile of Q may not match the CSP at the sequence level is true but irrelevant, since the match probability incorporates all sequence variants of the same length.

4.1.1.2   Stutter peaks

A *stutter peak* is a small non-allelic peak that appears on an epg adjacent to an allelic peak. At locus FGA in Figure 4.1c is a high peak labelled 25, beside which is a small, unlabelled peak at a location corresponding to an allele 4 bp shorter, that is, 24 repeat units. Since no other allele is detected at this locus, the possibility exists that the true underlying profile is a 24,25 heterozygote with unbalanced peaks. However, such an extreme peak imbalance is unlikely under normal conditions. In addition, the height of the peak at allele 25 supports the designation of the genotype as a 25,25 homozygote at this locus, in which case the small peak is a stutter artefact. The height of the peak is the most important feature of the signal, but some other aspects, such as its shape (or *morphology*) and area beneath the curve, can also be informative in deciding on such designations.

Stutter peaks are usually a PCR artefact, caused by mispairing of strands during replication. They usually have less than 10% of the height of the main peak, although the average height increases with allele length and can approach 15% for some long alleles (Butler, 2005). Because of this, stutter peaks are usually easy to distinguish from allelic peaks for a single-contributor sample. However, for mixed CSPs (Section 6.5) for which the different sources contribute to the sample in different proportions, the interpretation of stutter peaks can be problematic (Section 8.2.4).

Double stutter (a peak at two repeat units less than the allelic peak) and over-stutter (a peak at one repeat unit more than the allelic peak) are sometimes observed, with over stutter being particularly prevalent at the tri-nucleotide repeat locus, D22. The height of these peaks is usually at most 5% of that of the allele peak.

4.1.1.3 'Pull-up' peaks

These arise when an allele that is dyed with one colour also causes a peak to appear corresponding to one of the other dyes, which can be problematic if the location of the peak corresponds to an allele for each dye. A small pull-up peak can be seen in Figure 4.1c, directly below the large single peak at locus D21S11 of Figure 4.1b. Similarly, a pull-up peak is seen in Figure 4.1b directly below the large single peak of allele 11 at locus D16S539 of Figure 4.1a. These are more difficult to distinguish from allelic peaks when the DNA is at low template.

4.1.1.4 Mutations

STR mutations can be classified into two major types:

- *Germ-Line (or meiotic) Mutations.* These occur in the process of transmitting an allele from parent to child and can cause the child's allele to differ from its parental type.

- *Somatic (or Mitotic) Mutations.* These occur within an individual, in any cell of the body, and principally arise during cell duplication (mitosis).

Germ-line mutations can be important for paternity and other relatedness testing, see Section 7.1.8. For identification, somatic mutations are potentially of some importance. Conceivably, a mutant type found in some but not all of an individual's cells could result in different profiles being recorded from different body tissues taken from the same individual or perhaps from samples of the same tissue taken at different times. This is very rare, except for mtDNA (see Section 4.2). Somatic mutations seem capable of causing only false exclusion errors, not false inclusions.

Another possibility is that three distinct alleles could be recorded at a locus from the DNA of one individual, due to a somatic mutation. This is similar to a copy-number variant (CNV), discussed below.

4.1.1.5   Copy number variants (CNV) and silent alleles

Some individuals have an entire extra chromosome ('trisomy'). This is usually fatal, except for trisomies involving chromosome 21 or the sex chromosomes. These often lead to severe disorders, although an extra X chromosome can be relatively benign. A more realistic concern for forensic applications are CNVs, in which a genomic region may be deleted or duplicated on a chromosome. In the latter case, three alleles may be amplified by a particular PCR primer even when the sample has only one contributor. If only one locus displays three peaks and the peaks at the other loci are balanced, then the possibility that there is only a single contributor should be considered, in addition to the possibility that there is a low-template second contributor. Buckleton et al. [2005] briefly surveyed data suggesting a rate of triallelic profiles of between 1 and 20 per 10 000 single-locus genotypes (excluding sex-chromosome trisomies, which are more common).

CNVs can generate single-contributor profiles with no more than two distinct alleles represented per locus, but with an unusual pattern of peak heights either because of a deletion (single peak but of a height similar to heterozygote alleles) or because of a duplication where two of the three alleles are the same, generating unbalanced peak heights. A deletion can be regarded as a silent (or *null*) allele, which can also be generated by a PCR failure at one allele, possibly because of a mutation in the primer sequence. Individual silent alleles may be detected through reduced height of the observed peak [Cowell et al., 2014]. See Butler [2005] for approaches used to minimize the problem of silent alleles, which is rare with modern STR typing techniques.

Undetected silent alleles cause no problem for identification, provided that each silent allele is consistently unamplified in repeat PCR assays. In that case, CSP and defendant profile will correctly be recorded as matching if the defendant is the true source of the crime scene DNA. This might be expected to occur if both crime scene and defendant samples were profiled in the same laboratory. Otherwise, slight differences in protocol could generate a silent allele in one laboratory that is non-silent in another laboratory. As noted above, this rarely occurs and seems capable only of causing false exclusion errors rather than false inclusions. The false exclusion error rate can be minimised by retyping one or both samples if the profiles match at all but one or two alleles.

Silent alleles can be problematic in relatedness testing as the sharing of a silent allele may go unrecorded, which can generate an apparent exclusion for parent–child pairs.

Silent alleles should not be confused with 'allelic drop-out', in which one of the two alleles at a heterozygous locus is unamplified, or so weakly amplified that it cannot reliably be reported, because of unusual experimental conditions such as low DNA copy number (Section 4.1.3).

4.1.2   Contamination

As techniques for collecting samples and extracting and amplifying DNA become more sensitive, the problem of contamination becomes potentially more serious. Our environment is to a large extent covered with DNA, often dispersed by breathing or touching, and when low-template CSPs are being profiled, it is common that some DNA is amplified from individuals not connected with the crime. It is often important to distinguish between contamination that is

1. by environmental DNA at the crime scene when the sample is collected, and

2. subsequent to sample collection, for example, in the laboratory.

In some contexts, only the latter is considered to be 'contamination'. However, they cannot be distinguished from the profiling results and so both need to be considered together in the analysis.

Many steps can be taken to reduce laboratory-based contamination, including profiling crime scene and reference samples in different laboratories, as well as the use of ventilation hoods, and wearing of face masks. The DNA profiles of all police and scientific staff involved in investigating the crime scene or handling the evidence are usually kept for reference in case of contamination by one of these professionals. Such precautions can greatly reduce, if not entirely eliminate, the possibility of contamination after evidence has been sealed.

Environmental DNA accumulated during sample deposition or collection complicates interpretation (because additional unprofiled contributors have to be allowed for) and weakens the value of evidence implicating a true contributor, but it is unlikely to cause a false inclusion unless there are many other contributors, in which case the evidence will be weak. The most dangerous form of contamination is between different evidence samples, from either the same or different crime scenes. An instance of the latter form of contamination occurred in the United Kingdom in 2012, when a man arrested for a minor offence in Devon was later charged with a rape in Manchester and held in custody for several months before he was released after it became clear that his DNA had been transferred to a sample from the Manchester crime scene by within-laboratory contamination, due to erroneous reuse of a plastic tray [Rennison, 2012].

4.1.3   Low-template DNA (LTDNA) profiling

LTDNA refers to STR profiling of samples containing minute quantities of DNA, possibly from just a few cells. These may arise, for example, when DNA is extracted from fingernail debris, fingerprints, gloves, cigarettes or rope. There is no precise distinction between LTDNA and standard STR profiling: LTDNA techniques modify standard laboratory procedures to increase the sensitivity of STR profiling, for example, by increasing the number of PCR cycles or reducing reaction volumes.

The sensitivity of LTDNA techniques means that stochastic effects are stronger than that for standard STR profiles. Allele peaks can be highly variable in height and unbalanced across the two alleles of a contributor. *Drop-out* of individual alleles or even all alleles at a locus can occur, in addition to *drop-in* of sporadic alleles. If the DNA was exposed to the environment for some time before it was collected, and particularly if the environment was warm and/or humid, degradation effects may be important. Contamination is another problem that is magnified in LTDNA work, since few crime samples are not affected by any environmental DNA. Formulation of hypotheses and the calculation of likelihood ratios (LRs) for LTDNA profiles are discussed in Chapter 8.

4.1.3.1   DNA transfer

As the sensitivity of LTDNA profiling has increased, providing a boon to crime detection, thanks to the increased possibilities for obtaining DNA profiles from a crime scene, so has the danger of detecting DNA from a person unrelated to the crime on a crime scene item. Several transfer mechanisms have been confirmed experimentally, including direct (or *primary*) transfer through contactless mechanisms such as coughing or talking [Port et al., 2006], and secondary or sometimes tertiary transfer that can include a chain of contacts [Goray et al., 2012; Lowe et al., 2002]. Typically, the transferred DNA cannot be attributed to a specific body tissue, and is sometimes called 'trace DNA'; see Meakin and Jamieson [2013] for a review. DNA can also be transferred across items of clothing that are washed together or from one part of an item to another during transport, such as between handle and blade of a knife. There is a wide variability between individuals in their propensity to 'shed' DNA on contact, making it difficult to draw inferences about the mechanism or timing of transfer from the quantity and/or quality of DNA. Evidence on transfer mechanisms is frequently sought by courts, whereas current understanding and experimental evidence base are limited. This can lead to the dangers of giving personal experience undue weight in court (see Sections 11.4.5 and 11.4.6).

We suggest that when an opinion requiring subjective judgments based on personal experience is requested, it is useful to structure that opinion as much as possible, for example, using Bayesian networks (Section 7.2.3). Formalising the reasoning can allow the roles of firm scientific evidence and subjective opinion to be explicit and open to scrutiny and challenge and can pave the way for sensitivity analyses to understand the importance of different components of the reasoning. See Aitken et al. [2003] for an example of how to formulate such a Bayesian network in a transfer scenario.

4.2   mtDNA typing

The mtDNA is a circular molecule of about 16.5 kilobases (kb) in length – minuscule in comparison with the autosomes, which are typically around 10^5 kb in humans. Unlike nuclear DNA, mtDNA resides outside the nucleus of the cell and is carried by cytoplasmic energy-generating organelles called mitochondria. While the nuclear genome exists in two copies per cell, one paternal and one maternal in origin, mtDNA exists in multiple copies – the copy number varies widely by cell type, depending on the cell's energy requirements.

mtDNA samples have been widely used to infer aspects of human female population histories [reviewed in Jobling et al., 2004]. One of the key advantages of mtDNA over nuclear DNA for such work is its higher mutation rate which generates substantial diversity. Also, because it exists in multiple copies, mtDNA is easier to type from small and/or degraded samples, including ancient DNA samples, which also makes them useful for forensic identification [Tully et al., 2001]. They are widely used for samples containing little or no nuclear DNA, for example, shed hair, bone and burnt or other seriously degraded remains.

mtDNA is the only DNA type for which sequencing is currently the usual method of profiling. It is almost entirely maternally inherited and so not affected by recombination. An mtDNA sequence must, therefore, be analysed as a single allele, which makes it a poor discriminator between individuals that are even distantly related via their maternal lineages [Birky, 2001].

The non-coding 'control region' of approximately 1.1 kb includes two hypervariable regions, each of about 300 bp, having the greatest variability, which were previously the focus of interest for both population genetic and forensic work, but now sequencing the whole mtDNA genome is more common.

One consequence of the high mtDNA mutation rate is heteroplasmy: the existence within an individual of multiple mtDNA types (differing at one or several sites). This is essentially the same phenomenon as the autosomal somatic mutations discussed in Section 4.1.1, but mtDNA heteroplasmy occurs more frequently and is often associated with disease. It can be inherited, in which case it affects all the cells of the body in the same way. Provided that mtDNA heteroplasmy can be reliably identified, it increases the discriminating power of mtDNA profiles. However, different typing procedures have different abilities to detect and record heteroplasmy and this could potentially lead to false exclusion errors. See Tully et al. [2001] for a further discussion on heteroplasmy and its effects.

4.3   Y-chromosome markers

The Y chromosome is relatively short for a nuclear chromosome (60 000 kb) but still much longer than the mtDNA chromosome. For the most part, it is inherited uniparentally, through the paternal line, although short, terminal sections of the Y recombine with the X chromosome (these 'pseudo-autosomal' regions of the Y are not discussed further here). Unlike mtDNA which is carried by both sexes, Y chromosomes are carried only by males. Also in contrast with mtDNA, the Y chromosome is gene poor, but it does carry the sex-determining SRY gene.

Some particular uses of Y-chromosome profiles in forensic work include identifying the male contributor to a mixture when the major component is from a female, or when the male donor is aspermic and so sperm-separation techniques cannot be employed. Y-chromosome profiles are also useful in distinguishing multiple male contributors to a mixture. As for the autosomes, STRs form the predominant marker type for Y-chromosome forensic work, although SNP and other polymorphisms are available.

Information about Y-chromosome STR haplotype population frequencies is available at www.ystr.org. In rare cases, a supposedly Y-specific primer may amplify a homologous X sequence. In addition, duplication and triplication polymorphisms are relatively common on the Y chromosome, so that multiple alleles may occasionally be observed.

4.4   X-chromosome markers †

The X chromosome is, similar to the Y, involved in sex determination: girls receive an X from their father, boys receive a Y; both receive an X from their mother. Thus, unlike the Y, the X is not sex specific. At 165 000 kb, the X chromosome is much longer than the Y. Although still relatively gene poor compared with the autosomes, it has a much higher gene density than the Y.

Many diseases are caused by a defective X allele that causes no adverse effect when paired with a normal allele in females, while males are usually adversely affected. This distinctive mode of inheritance of diseases is referred to as 'sex linked'. The most common sex-linked diseases are red-green colour blindness (which affects 8% of men but only 0.5% of women), haemophilia A and several forms of mental retardation, including fragile-X syndrome.

X-chromosome STRs are not as widely used in forensic applications as those described above, but they do have some distinct advantages in paternity testing of girls. Because a man has only one allele to transmit to his daughter, there is never ambiguity about the father's transmitted allele, so that X-chromosome markers are, on average, more informative than autosomal markers. More importantly, if the girl's father is absent but her mother is available, the father's X-chromosome can be inferred and compared with the X chromosomes of close relatives of a putative father.

We will not discuss X-chromosome markers any further in this book. See Szibor et al. [2003] for discussion on their forensic uses and some basic statistical analyses and Ayres and Powley [2005] for inclusion probability and paternity index calculations.

4.5   SNP profiles †

The single-nucleotide polymorphism (SNP, pronounced 'snip') has become the marker of choice for many human genetic studies, because

- it is the most abundant type of polymorphism in the human genome;

- SNPs can be typed accurately, cheaply and in large volumes.

SNPs are predominantly diallelic, and so it is meaningful to speak of the 'minor' allele. SNPs with a relatively common minor allele (say, relative frequency > 10%)

are the most informative and are usually preferentially chosen for typing over SNPs with a rare minor allele.

It is natural to consider whether SNPs might make a more suitable typing system for forensic purposes than the STRs that dominate current forensic work. Although a typical STR locus is as informative as 4–6 SNPs, the two abovementioned advantages can outweigh this disadvantage. A bank of 50 SNPs will typically have at least the same information content as a dozen STRs [Sanchez et al., 2006]. The discriminatory power can also be increased by using triallelic SNPs [Westen et al., 2009]. Some more advantages of SNPs that are specific to forensic applications include:

- Improved typing of degraded and/or low-template samples due to shorter PCR amplicons than STRs ($s \leq 50$nt compared to $100 < s \leq 360$ nt for SGM+ loci).

- Lack of stutter peaks, due to non-repetitive nature of SNP markers.

Between Krjutškov et al. [2009] and Gill et al. [2004], four hindrances to widespread utilisation of SNP markers for forensic identification are highlighted:

Difficulty in Mixture Analysis. SNPs are relatively poorly informative about contributors to mixtures, since at each SNP, all combinations of genotypes involving at least one copy of each allele cannot usually be distinguished, yet this will be the norm.

Non-Compatibility of National STR Databases and SNP Data. Huge investment has already been made in accumulating large STR databases, for example, of previous offenders and unsolved crimes. Retyping with SNPs would be expensive even if the samples had been kept to make it feasible.

Large Amount of Input DNA Needed Relative to STRs. Krjutškov et al. [2009] stated that SNP typing requires approximately 15 times as much DNA as STR typing. While this figure may have changed since then, SNP typing still requires considerably more input DNA than STR typing.

Difficulty in Cost Saving. Although SNPs can be cheap to type in large volumes, for example, in the setting of genetic disease studies, the cost savings can be difficult to realise in forensic work which can involve small and degraded samples requiring individual attention.

Gill et al. [2004] also highlighted drop-out as a potential concern with SNPs, but statistical methods developed to deal with drop-out for STRs (See Chapter 8) should also be available for SNPs.

Kaur et al. [2014] have developed a regression framework for the interpretation of DNA mixtures using SNPs. In their method, no hypotheses are formulated, and the number of contributors does not need to be specified a priori; an individual is deemed to contribute to the sample if the regression coefficient for their contribution is greater than zero. Currently, this method can only deal with systematic signal strength differences (e.g. due to DNA template) and not with stochastic effects caused by degradation and/or low template. The model also does not include any profiled individuals, other than the queried individual; however, the authors state their intention to implement this feature. This work is in its early stages; however, it does demonstrate progress towards overcoming the problem of mixture deconvolution of SNP profiles.

We believe that with the continued difficulties posed by SNP typing in a forensic setting, they are now unlikely to become widely used in forensic genetics, but they will continue to be used for some specialist applications: mtDNA and Y-chromosome typing and to predict some aspects of ancestry or phenotype (Sections 7.4 and 7.5). SNPs also have advantages in analysing highly degraded samples that can arise from aircraft crashes and other disasters. See Kayser and de Knijff [2011] for a review of SNP typing for both forensic identification and phenotyping.

4.6   Sequencing †

While SNP typing on its own is unlikely to replace STR typing [Gill et al., 2004], full sequencing of STRs may become the standard in the future [Phillips et al., 2014], perhaps together with SNPs informative for ancestry and/or phenotype (see Sections 7.4 and 7.5). Sequencing of STRs can bring advantages in the detection of microvariants (Section 4.1.1), allowing for greater discrimination than is currently possible. Sequencing provides information about the depth of coverage that can help with the interpretation of mixed samples, much like epg peak height, and is similarly subject to variability due to extraneous factors [Phillips et al., 2014].

There are some obstacles that need to be overcome before sequencing can replace STR typing: large DNA inputs are required (1-10 ng), DNA preparation is labour intensive, alignment of repeat regions is difficult and there are not yet methods in place for the evaluation of evidential weight using sequence coverage. The biotechnology company Illumina has created a forensics division to tackle these problems.

4.7   Methylation †

Methylation refers to the process of adding a methyl group ($-CH_3$) to cytosine residues of DNA at the $5'$ position, particularly at CpG dinucleotides. Methylation has primarily been associated with gene silencing [Newell-Price et al., 2000], but there have been instances of DNA methylation being associated with activation of genes [Straussman et al., 2009]. Methylation, along with other epigenetic modifications, provides a mechanism for control of gene expression and permits adaptation to environmental changes over the lifetime of an organism [Szyf, 2011].

In a forensic setting, methylation patterns can potentially provide information on the tissue type of a sample, the age of the source individual, the time since deposition of the sample and the cause of death. Methylation studies have also been used to discriminate between monozygotic twins.

There are challenges to all of these potential uses of methylation. The dependence of methylation status at a given locus on many factors complicates inferences. These factors include disease status, particularly cancer [Wild and Flanagan, 2010];

environmental stress (e.g. from toxins) [Christensen et al., 2009]; ageing [Bocklandt et al., 2011; Koch and Wagner, 2011] and smoking [Christensen et al., 2009]. There are likely to be many other as yet unreported factors affecting methylation status. For more information on methylation in a forensic context, see Vidaki et al. [2013].

4.8   RNA †

RNA typing usually involves purifying the RNA in a sample to retain only those with a poly-A tail, indicating that they are messenger RNA (mRNA), the product of gene expression. The information available from mRNA should be highly correlated with that from DNA methylation, as both indicate expression levels. Many of the potential forensic uses of mRNA are similar to those for methylation; tissue type/body fluid identification [Juusola and Ballantyne, 2007] and age of a stain [Bauer et al., 2003a]. Other uses are more distinct: determination of wound age, post-mortem interval [Bauer et al., 2003b], circumstances of death [Ikematsu et al., 2006] or health of cells/organs [Liu et al., 2004]. However, methylation typing can be performed on the same sample as DNA typing is performed on, whereas mRNA typing necessitates splitting the sample into one portion for DNA typing and the other for mRNA typing. Additionally, RNA is considerably less stable than DNA, due to degradation by ribonucleases; mRNAs last from minutes to days after deposition (depending on transcript) [Vennemann and Koppelkamm, 2010], contrasting with DNA being stable potentially for years after deposition. Moreover, stability of mRNA is severely diminished by exposure to certain environmental conditions, such as moisture [Setzer et al., 2008]. For these reasons, RNA typing may be less useful than methylation typing. For more information on forensic RNA typing, see Bauer [2007] and Vennemann and Koppelkamm [2010].

Forensic uses of methylation and RNA profiling are in their infancy, and currently, no methods have yet been developed for the evaluation of the weight of evidence.

4.9   Fingerprints †

We are mainly concerned with identification evidence that is directly measured from DNA sequences, but it seems useful to make some comments on, and comparisons with, fingerprint evidence, which in some respects is similar.

Although fingerprints are partly under genetic control, there is sufficient environmental variation in the process of foetal development that even identical twins have distinguishable fingerprints. This weaker correlation between relatives gives fingerprints an advantage over DNA profile evidence and is the basis of the widely cited claim that all fingerprints are unique.

It is impossible to prove any human characteristic to be distinct in each individual without checking every individual, which has not been done. The complexity of the variation makes the claim at least plausible, and it is further supported by the successful use of fingerprint evidence for over a century. Yet, although there have been a number of theoretical investigations, the suggestion that recorded fingerprints are unique has never been rigorously checked. See Kaye [2003] for a critique of an attempt to try to provide scientific back-up for the claim.

Whatever the truth of the uniqueness claim for true fingerprints, it is of little practical relevance. What is at issue in a court case involving fingerprint identification is not:

'Is this person's fingerprint unique?'

Given two people, it is probably true that an expert could always distinguish their fingerprints given careful enough measurements. Instead, in a practical identification scenario, the question may be

'Is this imperfect, possibly smudged, smeared or contaminated, finger mark taken from a crime scene enough to establish that the defendant and nobody else could have left it?'

The answer to that question will always depend on the quality of the mark recorded at the crime scene, even if the general proposition of the uniqueness of fingerprints is accepted.

To help answer this question, UK police in 1924 adopted a '16 matching points' standard for reporting a comparison of a crime scene mark with a print, with 10 matching ridge characteristics sufficing for subsequent marks at the same scene. This standard was accepted for over 70 years but was finally abandoned in 2001 after being shown to have little scientific basis. No specific alternative numerical standard has replaced it. Confidence in fingerprint evidence has eroded as a result of high-profile errors, such as that of the Madrid bombings suspect in the United States [Office of the US Inspector General, 2006] and in the experimental evidence of errors in studies [Ulery et al., 2010]. The report of the National Research Council [2009] was critical of unscientific aspects of fingerprint evidence in current practice.

DNA evidence is superior to fingerprint evidence in that uncertainty can be measured reasonably well, and given enough markers, a high level of certainty can usually be achieved. In fact, the LR framework used for DNA profile evidence is now seen as setting a standard for rigorous quantification of evidential weight that forensic scientists using other evidence types should seek to emulate. See Neumann et al. [2012] for an approach to compute LRs for fingerprint evidence.

5

Some population genetics for DNA evidence

5.1 A brief overview

5.1.1 Drift

If we sample 10 balls at random from a bag on several occasions, the number of green balls chosen may vary even though the contents of the bag have not changed. This is the chance variation due to sampling, sometimes called 'sampling error'. It is what the '±3%' that often accompanies opinion poll results is trying to measure.

Sampling error also applies if we sample alleles (gene variants) in different populations, but the relative frequencies of the alleles also vary across populations. In fact, the evolution of genes from generation to generation in a closed population is like repeated, *compounded*, sampling of balls from a bag. Suppose that there are initially 10 individuals in the population, 5 'green' and 5 'blue'. Sex is an unnecessary complication here, and we will assume that all 10 individuals are equally fit and are capable of morphogenesis to produce a large number of offspring of the same colour as themselves. However, the environment will only support 10 adults, and so only 10 of the offspring survive to reproduce. In the absence of selective advantage, the number of green individuals in the second generation will have a random distribution centred at 5 (in fact a Binomial(10,0.5) distribution, with mean 5 and standard deviation (SD) $\sqrt{10 \times 0.5 \times 0.5} \approx 1.6$).

Suppose that the actual number of green individuals in the second generation happens to be 7. The random process of reproduction occurs again, and the number of green individuals in the third generation has a Binomial(10,0.7) distribution, with mean 7 and SD $\sqrt{10 \times 0.7 \times 0.3} \approx 1.4$. The number of green individuals in

Weight-of-Evidence for Forensic DNA Profiles, Second Edition.
David J. Balding and Christopher D. Steele.
© 2015 John Wiley & Sons, Ltd. Published 2015 by John Wiley & Sons, Ltd.
Companion Website: www.wiley.com/go/balding/weight_of_evidence

the population will 'drift' away from its starting value, until eventually 'fixation' of greenness or blueness occurs: everyone in the population is either green or blue.

If there are many populations evolving in the same way and with the same starting configuration, then eventually about half of the populations will be all-green and about half will be all-blue. Figure 5.1a shows the counts of green individuals in six populations, each evolved for 20 generations from a starting configuration of 5 green and 5 blue individuals. Five of the populations reached fixation, of which three were fixed for the green allele.

Alleles that have reached fixation are of little interest for forensic identification. In real human populations, there are many factors that tend to counter, or modify, the effects of drift. First, we consider the effects of population size and then move on to consider other evolutionary processes: mutation, migration and selection. We do not focus here on either recombination or gene conversion which, while important evolutionary processes, do not play an important role in the forensic issues discussed in this book.

5.1.1.1   Population size

Human populations are usually much larger than 10 individuals. Fortunately, for us, under simple models, genetic drift behaves in approximately the same way for different population sizes but at a speed that is inversely proportional to population size. Figure 5.1b illustrates a scenario in which the allele frequency and population size are both 100 times their values in Figure 5.1a and the pattern of fixation is similar except that it occurs 100 times more slowly. For this reason, much of population genetics theory works with time scaled according to the population size: this way the same formulas can be applied to a population of any size.

The situation is more complicated because it is not the census population size that matters but the *effective* population size, which is usually much smaller and is difficult to measure. The effective population size is the size of a hypothetical random-mating population that displays the same level of genetic variation as the observed population: it is the size of the theoretical population that best fits the actual population. There are several ways to define 'best fitting', and hence, several different definitions of effective population size. Note that there may be no random-mating population that fits the actual population well: the concept of effective population size is useful but not universally applicable.

Surprisingly, the effective size of the entire human population is often estimated to be about 10 000 adults, many orders of magnitude less than the current census size of several billions. There are many possible reasons for this extremely low value, among them the effects of rapid population growth, geographic dispersal, locally fluctuating population sizes, high between-male variance in reproductive success and inherited female fecundity. See Jobling et al. [2004] for a more extensive discussion. Remember that effective population size is a theoretical concept, and an effective population size of 10 000 does not imply that an ancestral population of size 10 000 had any special role in human history. It does mean that the actual human genetic variation looks in some respects similar to the variation that would be expected at neutral loci in a random-mating population of constant size 10 000.

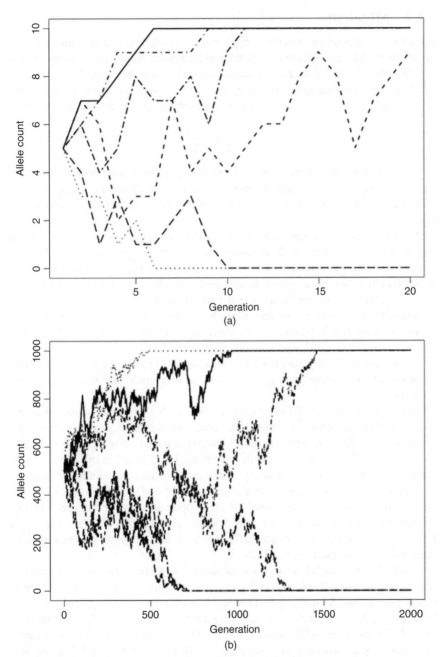

Figure 5.1   Frequencies over time of 'green' individuals in six haploid popula-tions. (a) Population size = 10, initial frequency = 5 and number of generations simulated = 20. (b) Population size = 1000, initial frequency = 500 and number of generations simulated = 2000.

5.1.2 Mutation

Mutation is the primary generator of genetic variation. Since drift tends to reduce genetic variation, an equilibrium can arise under neutral population genetics models in which the genome-wide average amount of variation is maintained at the level at which the new variants being created by mutation are balanced by the variants being lost through drift. Possibly, the most important formula in population genetics is

$$F = \frac{1}{1 + 4N\mu}, \tag{5.1}$$

where F denotes the probability that two chromosomes have had no mutation since their most recent common ancestor at a locus with mutation rate μ, under mutation-drift equilibrium in an unstructured, random-mating population of $2N$ chromosomes.

Most of the human genome has a very low mutation rate: for a typical nucleotide site, $\mu \approx 2 \times 10^{-8}$, or about 2 mutations per 100 million generations, and consequently, at mutation-drift equilibrium, there is relatively little variation. Indeed, the probability that two homologous human chromosomes chosen at random (worldwide) match at a given nucleotide site is typically about 0.9992. Although a match does not imply no mutation, because of the low mutation rate the two are expected to be almost equivalent in practice, and substituting $F = 0.9992$ and $\mu = 2 \times 10^{-8}$ into (5.1) leads to $N \approx 10\ 000$, as discussed in Section 5.1.1.

The control region of mitochondrial DNA (mtDNA) (Section 4.2) has a higher mutation rate, on average perhaps two orders of magnitude higher than autosomal nucleotides.

Short tandem repeat (STR) mutation rates are even higher (around 2 mutations per locus per 1000 generations). Perhaps the most common theoretical model for STR mutation is the stepwise mutation model (SMM) in which a mutant allele has either $k - 1$ or $k + 1$ repeat units, each with probability 1/2, where k is the current repeat number. The SMM has no stationary distribution, so that two similar populations isolated from each other do not converge under the SMM to the same allele frequency distribution. Thus, substantial between-population diversity is expected under the SMM for populations that exchange few migrants. In contrast, there is typically little between-population variation at human STR loci, suggesting high migration rates and/or the invalidity of the SMM.

In fact, the strict SMM is known to be false, for example, because the mutation rate increases with allele length and occasional two-step mutations occur, although it may provide an adequate approximation for some purposes. The SMM can easily be modified, for example, by hypothesizing a bias towards contraction mutations in long alleles, to obtain STR mutation models that do have a stationary distribution [Xu et al., 2000; Calabrese and Durrett, 2003; Whittaker et al., 2003; Lai and Sun, 2003]. Such models can provide a better fit to observed data than the SMM and are consistent with the observed between-population homogeneity of allele proportions at many STR loci.

5.1.3   Migration

From a global perspective, migration does not generate new variation, but human populations are geographically structured, and from a local perspective, migrants can be the most important source of genetic variation. Indeed, in simple models of a large population divided into subpopulations that exchange migrants, the formula describing the genetic diversity at a locus under migration-drift equilibrium exactly mimics the mutation-drift formula (5.1):

$$F = \frac{1}{1 + 4Nm},$$ (5.2)

where m is the proportion of subpopulation i that migrates to subpopulation j in each generation, assumed to be the same for all i and j and constant over time. Here, F is the probability that the two chromosomes have a most recent common ancestor within the subpopulation, without any intervening migration event.

The important implication of (5.2) is that relatively low levels of migration (but still much higher than the mutation rate, which is neglected) can suffice for F to be small. Substantial migration is a recurrent phenomenon throughout human history and much of prehistory [Jobling et al., 2004]. Thus, migration is thought to be the dominant factor in explaining the geographic distribution of human genetic variation. Indeed, genetic polymorphisms are used to trace historic migrations (see e.g. Romualdi et al., 2002; Hellenthal et al., 2014) and the effects of mutation often play little role in this work. At STR loci, the much higher mutation rate makes tracing migration events more difficult, other than that on the Y chromosome when multiple completely linked STRs can be exploited. Indeed, Rosenberg et al. [2002] found that nearly 95% of human genetic variation at autosomal STR loci was accounted for by within-population differences, and so many STR loci are required to distinguish human populations.

If the STR mutation process is at least approximately stationary, which is supported by the data and models cited in Section 5.1.2, then both migration and mutation tend to homogenise allele frequency distributions in different populations. There is a distinction in their effects in that a mutant STR allele tends to be similar in length to its parental type, whereas a migrant allele can in principle be of any allelic type. However, geographical structuring of human populations may mean that migrant alleles are also of a similar type to existing alleles in the subpopulation. Consequently, the effects of mutation are difficult to distinguish from those of migration in explaining current genetic variation.

5.1.4   Selection

Selection refers to the differential success of individuals according to their genetic make-up, where success is measured in terms of producing offspring and raising them to maturity. Some forms of selection can tend to reduce genetic variation within a (sub)population, such as the elimination of deleterious variants or the fixation of advantageous variants. Other processes, collectively referred to as balancing selection, tend to maintain genetic diversity, in some cases, at approximately the same level in different populations that are isolated from each other. The following are the most important forms of balancing selection:

Heterozygote Advantage. Individuals with two different alleles at a locus enjoy an advantage over homozygotes (this form of selection seems to be common in humans at loci involved in immune response); and

Frequency-Dependent Selection. For example, an individual of a type that is rare in the population has an advantage over those of common types (e.g. because it tends to escape the attention of predators).

A major controversy in the late 20th century population genetics was the extent to which observed genetic variation reflects the effects of selection versus those of drift. For a century after Darwin, it was widely accepted that most genetic variation had a selective explanation. The emergence of molecular variation data from the 1970s led to the neutral theory (e.g. see Kimura, 1977). Although hotly contested, the neutral theory has had a profound impact, so that for the past few decades, most population geneticists have assumed that most human genetic variation can be regarded as approximately neutral for most purposes. The huge amount of genomic data now becoming available has allowed this assumption to be checked more thoroughly: many loci show signs of the effects of selection, but still representing a relatively small fraction of the genome.

Some STR loci are directly implicated in causing human disease: an elongated coding-sequence CAG repeat causes Huntington disease, and other trinucleotide repeats peripheral to coding sequences can interrupt transcription, causing, for example, Fragile X syndrome. The tetranucleotide repeats typical in forensic applications are generally thought to have little or no phenotypic effect and, consequently, not to be under direct selection. However, there is little substantial evidence to support this, and some evidence to the contrary. Albanèse et al. [2001] reported an intronic tetranucleotide STR influencing transcription, and Subramanian et al. [2003] suggested a possible functional role for GATA repeats. Further, some long-established STR loci in forensic use are closely linked with disease-implicated genes (see Butler, 2005, and references therein), and hence, the patterns of variation at these loci are potentially affected indirectly by selection.

The selectionist–neutralist debate is not fully resolved, but it is clear that (effective) population size is crucial: the effects of drift are relatively more important when the population size is small. The low human effective population size means that humans have less genetic variation overall and are more susceptible to the effects of drift than might have been expected, given our large census population size. In the sequel, we will usually assume neutrality at the loci under consideration and, occasionally, comment on the possible effects of selection.

Selection almost certainly plays an important role in explaining the population variation at mtDNA and Y-chromosome markers (Sections 4.2 and 4.3), since for these chromosomes, there is no recombination to limit the extent of selective effects.

However, because few population-genetic assumptions are made in the interpretation of these markers, it may have little effect on their role for forensic work (outlined in Section 6.4).

5.2 F_{ST} or θ

Other factors, beyond those considered above, can modify the effects of drift, including admixture (e.g. when two previously distinct populations merge) and inbreeding. A detailed understanding of the influence of all factors on the evolution of profile proportions in human populations requires a lifetime of study and more. Although some understanding of the underlying processes is helpful, for the purposes of forensic applications, the most important question concerns the *magnitude* of the variation of allele proportions among different subpopulations, relative to the population from which a forensic database was drawn.

Each of the major populations can be regarded as being composed of many subpopulations, based, for example, on geography, ethnic identity or religion. For example, the population may correspond to 'UK Caucasians', and the subpopulations relevant to a particular crime may include, for example, people of Irish, Cornish, East Anglian, Orkney Island, Cypriot or Jewish ancestry. Because of migrations and inter-marriages, such groups are not strictly well defined, and simple population genetics models do not apply exactly. However, such models can form the basis for an analysis that attempts to assess the variation of allele proportions in such subgroups about a Caucasian average. The effects of this variation can then be allowed for in DNA profile match probability calculations, without a full knowledge of either the history of these groups or the population-genetic forces that caused the variation.

Wright [1951] introduced the coefficient F_{ST}, also called θ in forensic applications, which he interpreted as measuring the average progress of subpopulations towards fixation, and hence he called it a fixation index. $F_{ST} = 1$ implies that all subpopulations have reached fixation at the locus, possibly for different alleles in different subpopulations; $F_{ST} = 0$ implies that allele proportions are the same in all subpopulations, and so the population is homogeneous. F_{ST} can also be interpreted as measuring the relatedness among individuals within subpopulations relative to the total population. More relatedness within subpopulations, relative to that in the total population, means higher F_{ST} and a greater variation in allele fractions across subpopulations.

F_{ST} is defined in terms of the mean square error (MSE) of subpopulation allele proportions about a given reference value. If $\tilde{p}$ denotes the subpopulation proportion of an allele, and the reference value is p, then the MSE is the expected (or mean or average) value of $(\tilde{p} - p)^2$, and we have

$$\text{MSE}[\tilde{p}, p] = \text{E}[(\tilde{p} - p)^2] = F_{ST}p(1 - p). \tag{5.3}$$

In forensic applications, the reference value is typically the allele proportion in a large, heterogeneous population from which a database has been drawn. The implications of this are discussed further in Section 6.3.2.

Usually, in population genetics, the reference value p is the mean (either the average over all the populations or the expected value under some evolutionary model), in which case the MSE is the same as the variance. For example, if $p = 0.2$ and $F_{ST} = 1\%$, then the subpopulation allele proportions have mean 0.2 and SD $\sqrt{0.01 \times 0.2 \times 0.8} = 0.04$. The statistical rule-of-thumb 'plus or minus two SDs' then gives a rough 95% interval of 0.12–0.28 for a subpopulation proportion. This rule of thumb does not work well for p small, and we can compute better intervals using the beta distribution, introduced in Section 5.3.1. Here, the beta distribution gives almost the same 95% interval: $(0.13, 0.28)$.

If $\tilde{p}_1, \tilde{p}_2, \ldots, \tilde{p}_k$ denote estimates of $\tilde{p}$ in k different subpopulations, and they each have (known) expectation p and the same variance, then a natural estimator of this variance is

$$\mathrm{Var}[\tilde{p}] \approx \frac{1}{k} \sum_{j=1}^{k} (\tilde{p}_j - p)^2.$$

Substituting in (5.3) leads to the estimator

$$\widehat{F_{ST}} = \frac{\sum_{j=1}^{k} (\tilde{p}_j - p)^2}{kp(1 - p)}, \tag{5.4}$$

in which we introduce the notation $\widehat{}$ to denote an estimator. For example, if $p = 0.1$ and the allele proportions in three subpopulations are $0.05, 0.11$ and 0.13, then using (5.4), we would estimate

$$\widehat{F_{ST}} = \frac{(0.05 - 0.1)^2 + (0.11 - 0.1)^2 + (0.13 - 0.1)^2}{3 \times 0.1 \times 0.9} \approx 1.3\%.$$

Usually, p is not known exactly and must be replaced with an estimate in (5.4). Often, the estimate of p is the value that minimises $\widehat{F_{ST}}$, and it is customary to replace the k with $k - 1$ in (5.4) to compensate for the bias that arises from this choice. Here, if p were unknown the natural estimate would be $(0.05 + 0.11 + 0.13)/3 = 0.097$, which is close to 0.1 but leads to a much larger estimate $\widehat{F_{ST}} \approx 2\%$, because the bias correction has a large effect when k is small. In forensic settings, the estimate of p will usually be based on the allele proportion in the most relevant population database.

There are a number of problems with using (5.4) in practice, among them the fact that it assumes a common value of F_{ST} over subpopulations. For more sophisticated method-of-moments estimators, see Weir and Hill [2002], and for likelihood-based estimation, see Section 5.7. Steele et al. [2014b] provided extensive, worldwide, likelihood-based estimates of F_{ST}, discussed in Section 6.3.2.

5.2.1 Population genotype probabilities

An individual's genotype at an STR locus usually consists of two alleles, one paternal and one maternal in origin. Alleles are conventionally labelled according to the number of repeat units, so that a genotype might be represented by the unordered allele pair 7,9 (unordered because we usually cannot say which allele is paternal and which is maternal). If the individual has the same allele, say the 11, from both parents, the genotype would be represented as 11,11. Below, we will use A, B and C for arbitrary STR alleles, and we will write their population allele fractions as p_A, p_B and p_C.

These fractions are unknown and indeed unknowable because the relevant population is not well defined. However, population allele fractions are routinely estimated in several loosely defined populations, usually based on ethnic appearance as assessed by a police officer. In the United Kingdom, the most important ethnic groups for forensic DNA profiles are Caucasians, Afro-Caribbeans, South Asians (also called Indo-Pakistani), East/South-East Asians and Middle East/North Africans. Clearly, the classification of the UK population into these groups is arbitrary; many individuals do not fit well into any of them.

Allele fractions in these groups are estimated from samples whose size is typically a few hundreds: allowance for the effects of sampling error is briefly discussed in Section 6.3.1. However, in addition to the problem of defining the ethnic groups, the samples used to estimate allele fractions in them are not scientific, random samples but are 'convenience' samples whose representativeness is unknown. Figure 5.2 shows that the allele fractions do not differ dramatically over these groups, which is encouraging that the effects of ethnic misclassification and non-representative samples will typically not be great and that use of a sufficiently large F_{ST} value (Section 6.3.2) can compensate for these effects. Some exceptions do exist, such as the high allele fraction for allele 9.3 of TH01 in Caucasians (Figure 5.2).

5.3   A statistical model and sampling formula

5.3.1   Diallelic loci

5.3.1.1   The beta distribution

Between-population variation in allele proportions at a diallelic locus is often modelled by the *beta* distribution, which has probability density function[1] (pdf):

$$f(x) = cx^{\lambda p - 1}(1 - x)^{\lambda(1-p)-1},\tag{5.5}$$

where c is a normalising constant whose value is known but not needed here, $0 \leq x \leq 1$, and

$$\lambda = \frac{1}{F_{ST}} - 1.$$

The expectation and variance of the beta are, respectively, p and $F_{ST}p(1 - p)$.

Plots of the beta distribution for $F_{ST} = 1\%$, 2%, and 5% are shown in Figure 5.3. Each curve represents a theoretical distribution of the proportion of the 'green' allele in a subpopulation. For example, the solid curve in Figure 5.3c ($p = 20\%$, $F_{ST} = 1\%$) indicates that a subpopulation allele proportion is likely to be close to 0.2 and is almost

---

[1] Our parameterisation of the beta is not standard, but is convenient here. The standard parameterisation has $\alpha = \lambda p$ and $\beta = \lambda(1 - p)$.

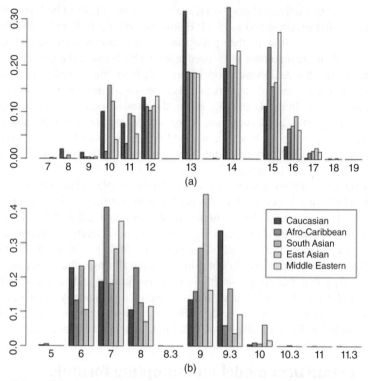

*Figure 5.2   Allele fractions in samples from five UK populations at two STR loci:
(a) D8 and (b) TH01. Sample sizes (alleles) are 6871, 3941, 520, 600 and 1202 for
D8 in population order shown in the legend box and 6816, 3918, 514, 598 and 1202
for TH01. See Steele et al. [2014b] for further details on the dataset.*

certainly between 0.1 and 0.3. More precise intervals can be evaluated by computing
areas under the beta density curve. This can be done easily in a statistical computer
package such as R (available free at www.r-project.org; use function pbeta)
or via a numerical approximation based on (5.5).

The beta distribution applies exactly under various theoretical models. It is unlikely
to be strictly valid in practice, but it usually provides a good approximation and allows
the essential features of genetic differentiation to be modelled and estimated in actual
populations.

5.3.1.2   A sampling formula for alleles

If the beta is adopted for the distribution of the allele proportion in a subpopulation, then a convenient recursive formula becomes available for the probabilities of
samples drawn at random from the subpopulation. This sampling formula for DNA
alleles can take the effect of relatedness into account. Suppose that *n* alleles have been

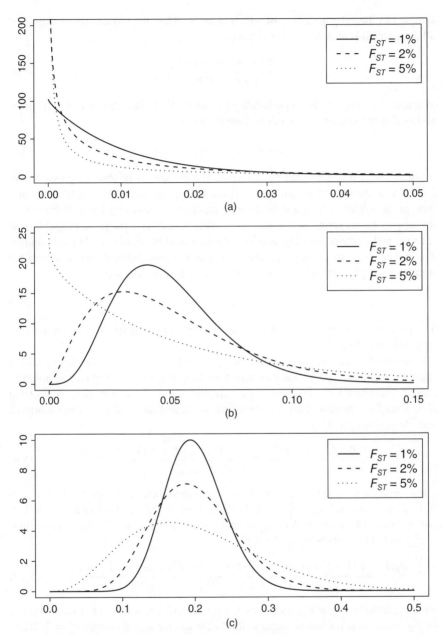

Figure 5.3   Probability density curves for the beta distribution when (a) p = 0.01 (1%), (b) p = 0.05 (5%) and (c) p = 0.20 (20%), and λ = 99, 49, and 19, so that F_{ST} = 1%, 2%, and 5%.

sampled in the subpopulation, of which m are A. Then the probability that the next allele sampled is also A can be written as

$$\frac{mF_{ST} + (1 - F_{ST})p_A}{1 + (n - 1)F_{ST}}.$$

(5.6)

When $m = n = 0$, we obtain probability p_A that the first allele drawn is A. The probability that the first two alleles drawn are both A is

$$p_A(F_{ST} + (1 - F_{ST})p_A) = p_A^2 + F_{ST}p_A(1 - p_A).$$

(5.7)

Roughly speaking, increasing F_{ST} increases the probability of observing two A alleles, because the first observation of an A allele suggests that they are relatively common in the subpopulation, and so drawing another A is less surprising. If there were no subpopulation variation, the second A allele would have the same probability as the first, and so the probability of two A alleles would be p_A^2, obtained by substituting $F_{ST} = 0$ in (5.7). Increasing F_{ST} also increases the probability of two B alleles but decreases the probability of an A allele followed by a B, which is

$$(1 - F_{ST})p_Ap_B.$$

(5.8)

The probability of an A and a B in an unordered sample of size two is obtained by multiplying (5.8) by 2.

Formula (5.6) can be used to build up probabilities for larger samples of alleles from a subpopulation. Samples of size four have a special importance in forensic identification problems, because of the two alleles at each locus from each of Q and X. Using (5.6), the probability of observing two A and two B alleles in an unordered sample of size four is

$$6\frac{p_Ap_B(1 - F_{ST})(F_{ST} + (1 - F_{ST})p_A)(F_{ST} + (1 - F_{ST})p_B)}{(1 + F_{ST})(1 + 2F_{ST})}.$$

(5.9)

For $p_A = 0.2$ and $p_B = 0.8$, (5.9) is shown as the dotted curve in Figure 5.4. At $F_{ST} = 0$, this curve has height $6p_A^2p_B^2 = 0.1536$; it rises slightly to 0.1540 at $F_{ST} = 1\%$, but subsequently starts to decline, reaching 0.1535 at $F_{ST} = 5\%$, and eventually 0 at $F_{ST} = 1$. The probability of AAAA is

$$\frac{p_A(F_{ST} + (1 - F_{ST})p_A)(2F_{ST} + (1 - F_{ST})p_A)(3F_{ST} + (1 - F_{ST})p_A)}{(1 + F_{ST})(1 + 2F_{ST})},$$

(5.10)

which approaches p_A as F_{ST} increases. For $p = 0.2$, (5.10) is shown as a dashed curve in Figure 5.4. It is the lowest curve in the figure when $F_{ST} = 0$ (height $p_A^4 = 0.0016$), but it increases markedly to 0.0041, when $F_{ST} = 5\%$, and eventually reaches 0.2 at $F_{ST} = 1$.

5.3.2   Multi-allelic loci

Formula (5.6) still holds if there are more than two alleles segregating at the locus. The multivariate extension of the beta distribution for the subpopulation allele proportions

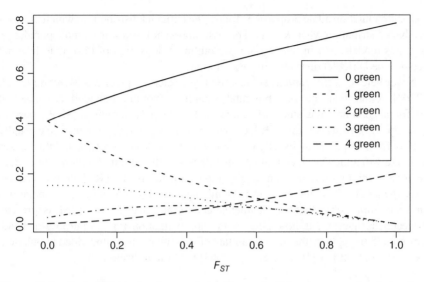

Figure 5.4   Probabilities for the number of 'green' alleles in a sample of size four under the beta-binomial sampling formula with $p = 0.2$ and F_{ST} ranging from 0 to 1.

is the Dirichlet, which has pdf

$$f(x_1, x_2, \ldots, x_K) = c \prod_{k=1}^{K} x_k^{\lambda p_k - 1}, \tag{5.11}$$

where $p_1, p_2, \ldots, p_K$ denote the population allele proportions, with $\sum_{k=1}^{K} p_k = 1$, and similarly, the x_k are all positive and sum to 1. If $K = 2$, then $p_2 = 1 - p_1$ and $x_2 = 1 - x_1$ and (5.5) is recovered.

Suppose that there are three alleles with population proportions p_1, p_2 and p_3, so that $p_1 + p_2 + p_3 = 1$. Successively using (5.6) and allowing for the six possible orderings, the probability P(1,1,1) that an unordered sample of size three from the subpopulation consists of one copy of each allele is

$$P(1, 1, 1) = \frac{6}{(1 - F_{ST})(1 + F_{ST})} \prod_{k=1}^{3} (1 - F_{ST}) p_k = 6 p_1 p_2 p_3 \frac{(1 - F_{ST})^2}{1 + F_{ST}}.$$

Similarly,

$$P(2, 1, 0) = 3 p_1 p_2 (1 - F_{ST}) \frac{(F_{ST} + (1 - F_{ST}) p_1)}{1 + F_{ST}},$$

$$P(3, 0, 0) = p_1 (F_{ST} + (1 - F_{ST}) p_1) \frac{(2 F_{ST} + (1 - F_{ST}) p_1)}{1 + F_{ST}}. \tag{5.12}$$

The latter two formulas are the same whether the locus is diallelic or multi-allelic.

The recursive sampling formula defined by (5.6) is very useful in population genetics, and especially for forensic applications. It has been derived under the assumption that subpopulation allele proportions follow the Dirichlet distribution (which includes the beta distribution when $K = 2$). This assumption holds in some simple population genetics models, and in Section 5.6, we outline a derivation of (5.6) that does not require the Dirichlet distributional assumption.

Nevertheless, (5.6) cannot be regarded as exact in practice. Marchini et al. [2004] found that the beta-binomial sampling formula provided an excellent fit for a genome-wide study of single-nucleotide polymorphism (SNP) markers (Section 4.5). However, for STR loci, mutation is such that the mutant allele usually differs from its parent by exactly one repeat unit (see Section 5.1.2), and this makes it unlikely that the Dirichlet assumption will be strictly valid if mutation is important relative to migration in explaining geographical patterns of STR allele proportions. More generally, the complex patterns of mating and migration of natural populations make it implausible that any mathematical formula can be regarded as precise. Instead, the question is how good is the approximation for the purpose at hand. Formula (5.6) captures the most important effects of drift in a subdivided population. See Weir and Ott [1997] and Balding [2003] for further discussion.

5.4   Hardy–Weinberg equilibrium

So far, we have been ignoring the fact that at autosomal loci, genes come in pairs, one maternal in origin, the other paternal. Ignoring this fact is justified under the assumption of HWE, which refers to the independence of an individual's two alleles at a locus. If HWE holds in an infinitely large population, then the genotype proportions in the homozygote and heterozygote case are of the form

$$
\begin{array}{lcc}
\text{Genotype:} & \text{AA} & \text{AB} \\
\text{Hardy–Weinberg proportion:} & p_A^2 & 2p_Bp_A
\end{array}
$$

HWE never holds exactly in real populations, but it provides a good approximation if the population size is large, if mating is at random in the population and if there is no differential survival of zygotes according to their genotype at the locus (i.e. no selection). If, however, the population is subdivided, then drift can cause it to deviate from HWE even if neutrality and random mating hold within subpopulations. Because an individual's two alleles form a sample of size two from a subpopulation, formula (5.6) applies (unordered case) and we obtain the genotype proportions

$$
\begin{array}{lcc}
\text{Genotype:} & \text{AA} & \text{AB} \\
\text{Proportion:} & p_A^2 + fp_A(1 - p_A) & 2(1 - f)p_Bp_A
\end{array}
$$

where we have followed the convention of using f $(= F_{IT})$ in place of F_{ST} when the two genes are drawn from the same individual, although $f = F_{ST}$ when, as we have assumed here, HWE applies within the subpopulations.

Inbreeding refers to a pattern of mate choice such that mates tend to be more closely related than random pairs of individuals within the population. Inbreeding in an unstructured population leads to genotype proportions exactly the same as those in a non-inbreeding, structured population given above, where f is now interpreted as the probability that the two genes uniting in a gamete are the same from a recent common ancestor. Indeed, the coancestry interpretation of F_{ST} (see Section 5.6) clarifies that population structuring and inbreeding have similar effects on individual genotype proportions. We briefly discuss the combined effects of both population structure and inbreeding on DNA profile match probabilities in Section 6.2.2.

Assortative mating refers to a practice of mate choice based on phenotype, for example, when mates tend to be more similar (e.g. in height, skin colour or intelligence) than are random pairs from the population. This affects genotype proportions in the same way as does inbreeding, except that it is limited to loci involved in determining the phenotype.

Other than population structure, inbreeding and assortative mating, the most important cause of deviations away from HWE is selection. In the extreme case of a lethal recessive allele, the homozygote genotype may be absent from the adult population even though the allele persists in heterozygote form.

In practice, HWE holds approximately in most human populations and at most loci, and often the reason for deviation from HWE in observed samples is genotyping errors rather than deviation from HWE in the underlying population. For example, the failure to observe one allele in a heterozygote (Section 4.1.1) may lead to the genotype being wrongly recorded as a homozygote for the other allele.

5.4.1   Testing for deviations from HWE †

We argue in Sections 5.4.2 and 6.2.1 that testing for deviations from HWE is not as important for forensic purposes as is often believed. However, it is, nevertheless, worthwhile, at least to check for genotyping or data recording errors, and so we briefly introduce some approaches to testing.

We describe and illustrate Pearson's χ^2 goodness-of-fit test and Fisher's exact test. Pearson's is the easiest test to apply, although Fisher's is usually superior in practice – see the discussion of Maiste and Weir [1995]. Often a better alternative is to estimate a parameter measuring divergence from HWE, and we also briefly introduce this approach below.

5.4.1.1   Pearson's Test

Consider the small dataset of genotypes at a diallelic locus shown in the 'observed' row of Table 5.1. The total allele counts are 25 A and 15 B, leading to an estimate of 25/40, or 5/8, for p_A. Using this estimate, the genotype probabilities *if* HWE holds are estimated to be

$$P(AA) \approx (5/8)^2 = 25/64$$

$$P(AB) \approx 2 \times (5/8) \times (3/8) = 30/64$$

$$P(BB) \approx (3/8)^2 = 9/64.$$

The corresponding expected counts in a sample of size 20 are then approximately 7.8, 9.4 and 2.8. These are not close to the observed counts but is the discrepancy large

Table 5.1   Pearson's test of HWE for a small dataset.

Genotype	AA	AB	BB	Total
Observed (O)	10	5	5	20
Expected (E)	7.8	9.4	2.8	20
$(O - E)^2/E$	0.61	2.04	1.70	4.36

enough that we should be convinced that HWE does not hold in the population from which the sample was drawn?

Pearson's goodness-of-fit statistic provides an answer to this question. The statistic can be computed as the sum of the squared differences between observed and expected values, each divided by the expected value (see Table 5.1). Alternatively, in the diallelic case, there is a shortcut formula:

$$n\left(\frac{ac - (b/2)^2}{(a + b/2)(b/2 + c)} \right)^2,$$

where a, b and c denote the counts of the AA, AB and BB genotypes, respectively. Here, $a = 10$, $b = c = 5$, and we obtain

$$20 \times \left(\frac{50 - 6.25}{12.5 \times 7.5} \right)^2 = 4.36,$$

which is greater than 3.84, the 95% point of the χ_1^2 distribution, but less than 6.63, the 99% point. Thus at the 5% level of significance, we can conclude that HWE does not hold, but we cannot reject HWE at the 1% significance level.

For a K-allele locus, the number of degrees of freedom of the χ^2 distribution is the number of heterozygote genotypes, which is $K(K - 1)/2$. The expected value of a χ_m^2 distribution is m, and so if the test statistic is less than m, we know without looking up tables that the null hypothesis cannot be rejected. The χ^2 distribution for Pearson's statistic is only approximate, and the approximation is poor if some of the expected genotype counts are small. As a rule of thumb, most of the expected counts should exceed five, and all should exceed two.

5.4.1.2   Fisher's exact test

Fisher's exact test of deviation from HWE is typically more powerful than the Pearson test and does not rely on the χ^2 approximation.

Consider a diallelic locus at which the genotype counts observed in a sample of size n are n_{AA}, n_{AB} and n_{BB}. Write n_A and n_B for the allele counts, and p_A and p_B for population allele proportions. Under HWE, the probability of the genotype counts *given* the allele counts is

$$P(n_{AA}, n_{AB}, n_{BB} \mid n_A, n_B) = \frac{n! \, n_A! \, n_B! \, 2^{n_{AB}}}{(2n)! \, n_{AA}! \, n_{AB}! \, n_{BB}!}. \tag{5.13}$$

(Remember that $n! \equiv n \times (n - 1) \times (n - 2) \times \cdots \times 2$, and $0! = 1! = 1$).

Table 5.2   Calculations for Fisher's exact test for the dataset of Table 5.1.

n_{AA}	n_{AB}	n_{BB}	Probability	Contribution to p-value
12	1	7	0.0004	0.0004
11	3	6	0.0028	0.0028
10	5	5	0.0370	0.0185
9	7	4	0.1764	
8	9	3	0.3527	
7	11	2	0.3078	
6	13	1	0.1105	
5	15	0	0.0126	0.0126
			1.0000	0.0343

Fisher's exact test uses the probability (5.13) as the test statistic. Consider all the possible ways of reassigning the observed alleles into genotypes. For example, if we observed $n = 2$ individuals with genotypes AA and BB, then there are just two possible genotype assignments with probabilities:

Genotypes	Probability
AA, BB	$(2!2!2!2^0)/(4!1!0!1!) = 1/3$
AB, AB	$(2!2!2!2^2)/(4!0!2!0!) = 2/3$

The p-value of the test is the total probability of all genotype assignments that are as or less probable, according to (5.13), than the observed assignment.

For the example introduced in Table 5.1, there are just eight possible genotype assignments possible given 25 A and 15 B alleles, and these are shown in Table 5.2. The genotype assignments with 1, 3 and 15 heterozygotes are each less likely than the observed assignment, and they have total probability 0.016. In small testing problems, it is customary to halve the probability assigned to the observed value – this is not necessary in practice because the probability of any particular observation is usually very small. Here, it leads to a p-value of just under 0.035, similar to that of Pearson's test.

5.4.1.3   Estimating the inbreeding parameter

To keep the discussion simple, we will here restrict attention to a diallelic locus and assume $p_A = 1 - p_B$ to be known. See Ayres and Balding [1998] for Bayesian estimation of f allowing for uncertainty in the population allele proportions and for multi-allelic extensions.

Under the inbreeding model, the homozygote and heterozygote probabilities are

$$P(AA) = p_A^2 + f p_B p_A$$
$$P(AB) = 2(1 - f) p_B p_A$$
$$P(BB) = p_B^2 + f p_B p_A,$$

where $\max(-p_B/p_A, -p_A/p_B) \leq f \leq 1$. Under this model, the likelihood of a sample with genotype counts n_{AA}, n_{AB} and n_{BB} is

$$L(f) = c P(AA)^{n_{AA}} P(AB)^{n_{AB}} P(BB)^{n_{BB}},$$

where c is an arbitrary constant. One possibility is to choose c such that $L(f)$ takes value 1 at the HWE value $f = 0$, in which case we obtain

$$L(f) = \left(1 + \frac{f p_B}{p_A}\right)^{n_{AA}} (1 - f)^{n_{AB}} \left(1 + \frac{f p_A}{p_B}\right)^{n_{BB}}.$$

If instead c is chosen so that the integral over f is 1, then the likelihood also specifies the posterior density curve for f given a uniform prior distribution. Figure 5.5 plots this density curve for the data of Table 5.1, assuming p_A and p_B equal to their sample proportions.

Bayesian hypothesis testing is usually performed via the Bayes' factor, which is the probability of the observed data under the inbreeding model divided by its probability under HWE. Here the Bayes' factor turns out to be about 2.7 in favour of

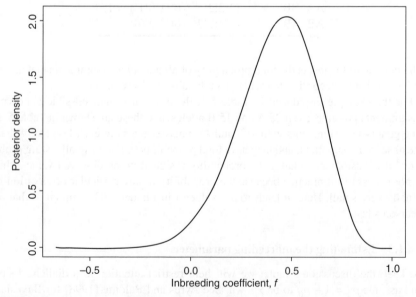

Figure 5.5   Posterior probability density for the inbreeding coefficient f, given sample genotype counts $n_{AA} = 10$, $n_{AB} = 5$ and $n_{BB} = 5$ and $p_A = 0.6$, $p_B = 0.4$ and a uniform prior.

the inbreeding model, but this depends sensitively on the choice of a uniform prior distribution for f. An informal Bayesian hypothesis test, less sensitive to the prior, can be achieved by looking at whether or not the HWE value falls within a given interval of highest posterior density for f. Here, the 95% highest posterior density interval is $(0.050, 0.784)$, excluding $f = 0$, but the 99% interval is $(-0.085, 0.849)$ which does include the HWE value. The results of this test are thus broadly in line with those of the Pearson's and Fisher's tests above, but in addition, we have Figure 5.5, which gives a visual representation of the values of f supported by the data. Because of the small sample size, these values span a wide interval, but the most likely value (the maximum-likelihood estimator) is $\hat{f} = 0.47$, suggesting a large deviation from HWE.

5.4.2   Interpretation of test results

Care must be taken with the interpretation of the results of any test, since the magnitude of any deviation of HWE is confounded with sample size: a false null hypothesis of HWE may well be accepted when the sample size is small. Conversely, HWE is never strictly true, and given a large enough sample size, it will be rejected even when the magnitude of the deviation from equilibrium is too small to be of any practical significance. An additional potential pitfall in hypothesis testing relates to the problem of multiple testing. If many loci are tested, by chance, some will indicate significant disequilibrium even if all loci are in fact in HWE.

It is widely believed that HWE in the population from which a population database is drawn is required for the validity of DNA profile match probability calculations using that database. This is incorrect. Both population subdivision and inbreeding increase the probability of shared ancestry (or 'coancestry') of the two genes at a locus and are one of the main potential causes of deviation from HWE. However, their effect on forensic match probabilities can be accounted for using F_{ST}, as discussed in Chapter 6. Deviation from HWE due to assortative mating and/or selection is expected to be small at forensic loci; even if there is some selection effect due to linkage disequilibrium with a functional locus or regulatory sequence, it is likely to be limited to one or two loci of the DNA profile.

Consequently, testing for HWE is not crucial for forensic work. A HWE test is easy to perform and can signal genotyping or data entry errors and so is probably worthwhile as a routine check. For forensic applications, accepting the null hypothesis of HWE is no reason for complacency, because we are concerned with the joint probabilities of four alleles, not just two. Conversely, rejecting the null is not necessarily a cause for concern, since this may represent non-HWE due to population subdivision, which is accounted for in the likelihood ratio (LR) via F_{ST}.

5.5   Linkage equilibrium

Linkage equilibrium (or gametic equilibrium) is the term used in population genetics for the independence of the alleles at different loci in the same gamete[2]

---

[2] That is, inherited from the same parent.

at distinct loci. There are many possible causes of linkage disequilibrium (LD). For non-forensic applications, the most important cause is linkage: the tendency for alleles at loci close together on a chromosome to be passed on together over many generations, because recombinations between them rarely occur. In a simple, deterministic, population-genetic model, the LD between two loci decreases exponentially over time at a rate proportional to the recombination fraction. Although often cited in textbooks, in practice, there is typically so much stochastic noise that the exponential pattern of decay is hardly perceptible. The LD in actual human populations usually extends over a few tens of kilobases. Occasionally, it extends over several hundred kilobases, but even this is small compared with a typical chromosome length of around a hundred thousand kilobases. Marker loci used in forensic work are chosen sufficiently far apart that LD due to linkage is negligible for unrelated individuals, except for multiple markers from the Y or the mitochondrial chromosomes, for which recombination is (almost) absent. Linkage can be important when considering relatives of Q as alternative contributors of DNA (Section 7.1.4).

Other than linkage, LD can be caused by population subdivision and drift. If a population is subdivided, drift may cause the allele proportions at many loci to differ from the population proportions, leading to statistical associations between unlinked loci. Another possible cause is admixture, in which a hybrid population is formed by migrants from several ancestral populations. For both subdivision and admixture, an individual with an allele at one locus that is frequent in one of the subpopulations (or source populations) may have a high level of ancestry from that (sub)population, and hence, at a second locus, she/he may also have an allele common in that (sub)population.

In forensic match probabilities, LD due to subdivision and drift is accounted for via F_{ST}. LD due to admixture is slightly different, in that an admixed population need not have any subdivision, and so the definition of F_{ST} in terms of subpopulation allele proportions does not apply. However, while subdivision and drift can sustain LD between unlinked loci, this is not true for a random-mating hybrid population, in which LD decays markedly in a small number of generations. Moreover, if a defendant has coancestry with some of the alternative possible culprits in a migrant gene pool, the interpretation of F_{ST} as the probability that two alleles are identical by descent (IBD, see Section 5.6) clarifies that it can also be used to adjust for this possibility.

Selection is another possible cause for LD at unlinked loci. Whatever the cause of LD, it only matters for forensic applications if it tends to enhance match probabilities. If selection is environment specific and, hence, tends to vary with geography, its effect is similar to that of drift and can be accounted for using F_{ST}. Other forms of selection are not expected to systematically affect forensic match probabilities.

The principles for, and caveats associated with, testing for deviations from HWE also apply to testing for LD. The tests differ according to whether genotype or haplotype data are available. Multi-locus genotype data consists of two alleles at each locus. A haplotype includes one allele from each locus, these alleles having the same gamete (i.e. parent) of origin. Each genotype thus corresponds to two haplotypes.

Consider first haplotype data at two diallelic loci, with alleles B and b, and A and a, respectively. Pearson's statistic for testing the hypothesis of linkage equilibrium is nr^2, where r denotes the sample correlation coefficient defined by

$$r = \frac{ad - bc}{\sqrt{(a + b)(c + d)(a + c)(b + d)}},$$

and a, b, c and d denote the sample counts of AB, Ab, aB and ab haplotypes, respectively. Sample sizes are usually sufficient to test for independence at only two or three loci using Pearson's test. See Zaykin et al. [1995] for a discussion of exact tests.

For testing deviations from LD using genotype data, see Weir and Ott [1997] and Schaid [2004].

5.6   Coancestry †

In this section, we derive the sampling formula (5.6) in a simple population genetics model without explicitly invoking the beta or Dirichlet distributions. It follows that although F_{ST} was introduced above in terms of the variation in subpopulation allele proportions, it can also be interpreted as the probability that two alleles are descended identically from a common ancestor. Hence, F_{ST} is also called a 'coancestry coefficient' or 'kinship coefficient'.

We adopt a simple model in which the population size is a large, constant N (alleles), and in each generation, each allele has probability u/N that it has arisen as a novel draw from a gene pool such that it is 'green' with probability p. Otherwise, it is an exact copy of one of the N alleles in the previous generation, each equally likely to be the parent. 'Draw from the gene pool' can be thought of as representing either mutation or migration, or both. Under this model, we outline a derivation of particular cases of the sampling formula for samples of size up to three and then give a general recursive argument.

5.6.1   One allele

In tracing the ancestry of an allele backwards in time, eventually, its ancestor will have been drawn from the gene pool, and hence, the probability that it is green is $P(1, 0) = p$.

5.6.2   Two alleles

Tracing the ancestry of two alleles backward in time, in each generation, there is probability $1/N$ that they have a common ancestor and probability $2u/N$ that one of them is selected from the gene pool. Thus, the probability that the two alleles are IBD from a common ancestor in any generation is $1/(1 + 2u)$, and the probability that this ancestor was green is again p. Reasoning similarly for the probability that the two alleles were both drawn at random from the gene pool and are both green and summing these two terms give

$$P(2,0) = \frac{p + 2up^2}{1 + 2u}$$

for the overall probability that both alleles are green which, on replacing $1/(1 + 2u)$ with F_{ST}, is the same as (5.7).

5.6.3   Three alleles

The probability that two of the alleles meet in a common ancestor without any of them having been drawn from the gene pool is

$$\frac{3/N}{3/N + 3u/N} = \frac{1}{1 + u} = \frac{2F_{ST}}{1 + F_{ST}}.$$

Continuing back in time, the probability that the remaining two alleles also meet in a common ancestor is F_{ST}, and so the probability that all three alleles are IBD is

$$\frac{2F_{ST}^2}{1 + F_{ST}}.$$

Similarly, it can be seen that the probabilities of exactly one and zero pairs of alleles being IBD from a common ancestor in the subpopulation are

$$\frac{3F_{ST}(1 - F_{ST})}{1 + F_{ST}} \quad \text{and} \quad \frac{(1 - F_{ST})^2}{1 + F_{ST}}.$$

The probability that three alleles sampled are all-green is then

$$P(3,0,0) = p\frac{2F_{ST}^2}{1 + F_{ST}} + p^2\frac{3F_{ST}(1 - F_{ST})}{1 + F_{ST}} + p^3\frac{(1 - F_{ST})^2}{1 + F_{ST}},$$

which simplifies to give (5.12).

5.6.4   General proof via recursion

Consider tracing back the ancestry of a sample of n alleles, of which m_1 are green and $m_2(= n - m_1)$ are not. With probability $(n - 1)/(n - 1 + 2F)$, the most recent event was a pair of alleles coalescing in a common ancestor, without any allele having been drawn from the gene pool. In this case, the current sample has m_1 green alleles if either

- the ancestral allele was green and was one of $m_1 - 1$ green alleles in the ancestral sample or

- the ancestral allele was non-green and the ancestral sample had m_1 green alleles.

Similarly, in the case that the most recent event was a draw from the gene pool, in order that the current sample have m_1 green alleles either

- the new draw was green and there were previously $m_1 - 1$ green alleles or

- the newly drawn allele was non-green, and there were previously m_1 green alleles.

Summing these four terms and substituting $F_{ST} = 1/(1 + 2F)$, we obtain the recursive equation:

$$P(m_1, m_2) = \frac{F_{ST}}{1 + (n - 2)F_{ST}}[(m_1 - 1)P(m_1 - 1, m_2) + (m_2 - 1)P(m_1, m_2 - 1)]$$

$$+ \frac{1 - F_{ST}}{1 + (n - 2)F_{ST}}[pP(m_1 - 1, m_2) + (1 - p)P(m_1, m_2 - 1)]$$

$$= \frac{(m_1 - 1)F_{ST} + (1 - F_{ST})p}{1 + (n - 2)F_{ST}}P(m_1 - 1, m_2)$$

$$+ \frac{(m_2 - 1)F_{ST} + (1 - F_{ST})(1 - p)}{1 + (n - 2)F_{ST}}P(m_1, m_2 - 1). \tag{5.14}$$

It is readily verified that the unordered form of (5.6) satisfies (5.14), provided that $P(i, j)$ is interpreted as zero if either $i < 0$ or $j < 0$.

5.7   Likelihood-based estimation of F_{ST} †

We briefly discussed a simple method-of-moments estimator of F_{ST} in Section 5.2. In this book, the sampling formula (5.6) will be mainly used to calculate LRs in Chapter 6, but we note here that it can also be used for likelihood-based estimation of F_{ST}.

First, to keep matters simple, we restrict attention to a single, diallelic locus and assume that the reference allele proportions are known. Specifically, we consider a sample of 10 alleles of which 6 are green, and suppose that $p_A = 0.2$. The sample proportion 6/10 is much larger than the expected proportion p_A, suggesting that F_{ST} is rather large, but any inferences must be weak with so little data. To make this precise, we successively use (5.6) to obtain

$$L(F_{ST}) = \frac{\prod_{i=0}^{5}(iF_{ST} + (1 - F_{ST})/5) \times \prod_{i=0}^{3}(iF_{ST} + 4(1 - F_{ST})/5)}{(1 - F_{ST})\prod_{i=1}^{8}(1 + iF_{ST})}$$

This curve is plotted in Figure 5.6 (solid line). As expected, a wide range of F_{ST} values are supported. The curve has been scaled so that it can be interpreted as a posterior density given a uniform prior for F_{ST}, and the 95% highest posterior density interval for F_{ST} runs from 0.027 to 0.83.

Figure 5.6 also shows the corresponding curve when the sample size is increased by a factor of 10, with the sample proportion of green alleles unchanged. Inferences are now a little stronger: the 95% highest posterior density interval is $0.099 \leq F_{ST} \leq 0.84$, excluding a larger interval near zero.

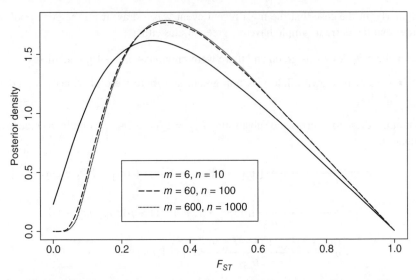

Figure 5.6   Likelihood curves for F_{ST} from a sample in one subpopulation, at a diallelic locus with $p = 0.2$. The curves have been scaled so that they can also be interpreted as posterior density curves given a uniform prior for F_{ST}.

Stepping up the sample size by a further factor of 10 (dotted curve), the posterior density curve is almost unchanged. This reflects the fact that it is only the subpopulation allele proportion that is informative about F_{ST}, and once the sample size is large enough for this to be estimated well, there is no additional benefit from increasing the sample size for that subpopulation and locus.

Instead of increasing the sample size at a particular locus/subpopulation, better inferences about F_{ST} can be obtained by combining information across loci and/or across subpopulations, but without assuming that F_{ST} is constant. BAYESFST is a C program for Bayesian estimation of F_{ST} from samples of STR alleles from several loci and several subpopulations that can be obtained from

www.rdg.ac.uk/statistics/genetics.

See Balding [2003] for the theory and Balding et al. [1996], Balding and Nichols [1997], Ayres et al. [2002] and Steele et al. [2014b] for applications.

5.8   Population genetics exercises

Solutions start on page 185.

5.1   The population proportion of an allele is $p = 0.15$. The population is subdivided and $F_{ST} = 2\%$ for each subpopulation.

(a) What is the SD of the allele proportion in the subpopulations?

(b) Find an approximate 95% interval for the subpopulation allele proportion (if possible, use the beta distribution, otherwise use a simple approximation).

5.2   You observe in large samples from five subpopulations the following proportions of an allele at a diallelic locus: 0.11, 0.15, 0.08, 0.09 and 0.12.

(a) You know that $p = 0.1$. Estimate the value of F_{ST} (assumed the same for each subpopulation).

(b) How would your answer from (a) be affected if p was unknown?

5.3   Use the sampling formula (5.6) to evaluate the probabilities of

$$\text{(a) } m = 3 \qquad \text{and} \qquad \text{(b) } m = 2.$$

A alleles in an unordered sample of size $n = 3$ (alleles) from a haploid population at a diallelic locus. Express your answers in terms of p (the population proportion of A) and F_{ST} and then evaluate them when $p = 0.25$ and $F_{ST} = 0\%$, 2% and 10%. (Hint: as a check, verify that you obtain 0.25 and 0 when $F_{ST} = 1$.)

5.4   Test for deviation from HWE in the population from which the following genotype proportions were observed at a triallelic locus:

Homozygotes:	AA 10	BB 5	CC 5
Heterozygotes:	AB 15	BC 10	AC 20

5.5   If the observed haplotype frequencies at two diallelic loci are

Haplotype:	AB	Ab	aB	ab
Frequency:	10	20	5	8

does linkage equilibrium hold in the population at these loci?

6

Inferences of identity

6.1 Choosing the hypotheses

We saw in Section 3.2 that the weight-of-evidence formula requires the likelihood ratio (LR) R_X for different individuals X other than the suspected contributor Q. An LR (introduced in Section 3.1) measures the weight of evidence in favour of one hypothesis relative to a rival hypothesis. It is not an 'absolute' measure of evidential weight but depends on how you formulate the competing hypotheses: there are often a number of different hypothesis pairs that might be considered, each with its own LR.

In a typical criminal case, the hypotheses of direct interest to the court concern the guilt or innocence of Q. In some cases, it may not be clear that a crime has occurred, and if so, whether it had a single perpetrator; but when these conditions do hold, the hypotheses of interest are

$$G \quad : \quad Q \text{ is the culprit;}$$

$$I \quad : \quad Q \text{ is not the culprit.}$$

The hypothesis pair (G, I) is clear and concise, but we noted in Section 3.5.1 that it is not a practical choice for quantitative evaluation of evidence: no LR can be directly computed without a further assumption allowing probabilities to be specified under hypothesis I. Instead, in Section 3.5.1, we discussed the partition of I according to different alternative contributors X.

In complex settings, such as those involving multiple contributors to the crime sample (see Section 6.5), the selection of hypotheses may be far from straightforward. In choosing hypotheses to consider, there is almost always a tension between

Weight-of-Evidence for Forensic DNA Profiles, Second Edition.
David J. Balding and Christopher D. Steele.
© 2015 John Wiley & Sons, Ltd. Published 2015 by John Wiley & Sons, Ltd.
Companion Website: www.wiley.com/go/balding/weight_of_evidence

allowing for all the realistic alternative possible explanations and trying to keep the formulation simple enough to be practical.

There can also be a tension between addressing the hypotheses most relevant to the court's task and respecting the boundary of the forensic scientist's domain of expertise. Commenting directly on the hypothesis pair (G, I) would under some legal systems expose a forensic scientist to the criticism that he/she has offended the 'ultimate issue rule'. This 'rule' seems not to have a precise statement in law but, in the context of a criminal trial, is generally interpreted as prohibiting an expert witness from giving a direct opinion about the guilt or innocence of Q. The concern motivating the rule seems well grounded: that the forensic scientist should not usurp the function of judge or jury, nor act as an advocate for the defendant. It can be difficult to demarcate the boundary between acceptable and unacceptable statements, and few legal authorities have attempted to do this [Robertson and Vignaux, 1995]. Therefore, an overly strict interpretation of the rule could diminish a forensic scientist's ability to give full assistance to jurors in their task of reaching a reasoned opinion on the ultimate issue (see Section 11.4.3). However, a reasonable interpretation of the rule accords with the principles that we have been advocating in this book and can be helpful. For a further discussion, see Robertson and Vignaux [1995].

Rather than addressing questions of guilt or innocence, it is usually more appropriate for a DNA expert forensic scientist to focus on questions of source attribution, such as

$$G' \quad : \quad Q \text{ is the source of the crime-scene DNA sample;}$$

$$I' \quad : \quad Q \text{ is not the source.}$$

In an allegation of rape, consent may be an issue. This would imply an important difference between G and G', yet the issue of consent would usually lie outside the forensic scientist's domain. Alternatively, the evidence may concern a stain found on the defendant's clothing that could have come from the victim, so that the relevant hypotheses might be

$$G'' \quad : \quad \text{the victim is the source of the DNA sample;}$$

$$I'' \quad : \quad \text{the victim is not the source.}$$

A victim's blood could be on the clothing of Q even though he/she did not cause the injury, in which case G and G'' are distinct.

Cook et al. [1998] and Evett et al. [2000] introduced hierarchies of hypotheses and discussed a rationale for choosing the appropriate level of the hierarchy. To keep the discussion and notation as simple as we can, where no confusion seems possible, we will continue to use (G, I) to denote the hypothesis pair of interest, even though (G', I') is more likely to be the actual hypothesis pair to be addressed by the forensic scientist. The issues are broadly the same if the actual hypothesis pair is (G'', I''), except that in the discussions of relatedness below we would be concerned with the relatedness to the victim of the alternative possible sources of DNA, rather than their relatedness to Q.

6.1.1   Post-data equivalence of hypotheses

Meester and Sjerps [2003, 2004] discussed the 'two-stain' problem in which a defendant's DNA profile matches that obtained from two distinct crime stains and different LRs can be obtained by contrasting different pairs of hypotheses. This causes no difficulty, provided that the weight-of-evidence formula (3.3) is used to combine the LRs.

More generally, Dawid [2001, 2004] noted that once the data have been observed, there can be sets of hypotheses pairs that were distinct a priori but which have been rendered equivalent by the data. The hypotheses pairs in such a set are said to be 'conditionally equivalent'. The pairs may have different LRs but, because they are conditionally equivalent, the allocation of posterior probabilities must, logically, be the same whichever pair of hypotheses is used for the analysis.

For example, let us return briefly to the island of Section 2.2.1, where life and crime are simple, thanks to a number of assumptions. Suppose that two suspects Q and Q' are investigated, and Q is found to have Υ while Q' does not. Consider the hypothesis pairs:

$$H_1 \quad : \quad Q \text{ is guilty;}$$

$$H_2 \quad : \quad Q \text{ is innocent;}$$

and

$$H_1' \quad : \quad \text{either } Q \text{ or } Q' \text{ is guilty;}$$

$$H_2' \quad : \quad \text{both } Q \text{ and } Q' \text{ are both innocent.}$$

These pairs become logically equivalent following the observation of the Υ states of Q and Q', given the fact that the culprit is a Υ-bearer. Under the assumptions of the island problem, the LR for H_1 vs H_2 is double that for H_1' versus H_2'. The prior probability of H_1 is half that of H_1' and, since these two effects cancel, the posterior probability of H_1 equals that of H_1'.

In this example, we obtain the same probability that Q is the culprit, whichever pair of hypotheses is contrasted, even though the LRs differ by a factor of two. Strictly speaking, either pair of hypotheses could be employed by a rational juror to arrive at the same conclusion. Nevertheless, although not a logical necessity, if Q is on trial, then the argument that (H_1, H_2) is the most appropriate hypothesis formulation seems compelling, as these hypotheses directly address the question of interest to the court. In contrast, the hypothesis pair (H_1', H_2'), which is conditionally equivalent to (H_1, H_2), is more complex for no obvious gain – it would invite confusion to introduce an LR for this hypothesis pair to measure the weight of evidence against Q.

Similarly, in simple identification scenarios with a single crime stain assumed to have originated from a single contributor, it is natural to compare G with all the hypotheses of the form 'X is the culprit', for different X. Then there is not one LR, but several, according to the differing levels of relatedness of X with Q. A collection of LRs for various levels of relatedness might make a useful summary of evidential weight. A single LR comparing Q with an unrelated X can make a useful summary when close relatives are unequivocally excluded, provided that the complexity of the underlying situation is appreciated.

When Q has been identified through having the only matching profile in an intelligence database of DNA profiles (Section 3.4.5), Stockmarr [1999] advocated contrasting the hypotheses

G^* : one of the persons whose profiles are in the database is the culprit;

I^* : not G.

Assuming no error, the pair (G^*, I^*) is equivalent to (G, I), given the data that Q gives the only match in the database. Therefore, it is possible to work coherently with these hypotheses and arrive at the appropriate probability of guilt. However, because (G^*, I^*) does not address the question of interest to the court, LRs related to these hypotheses can be seriously misleading if proposed as measures of evidential weight. For example, the LR based on (G^*, I^*) depends on database size, which can lead to an erroneous conclusion that the evidential strength of a DNA profile match weakens with increasing database size (see Section 3.4.5). For further discussion and criticisms of Stockmarr [1999], see Donnelly and Friedman [1999], Balding [2002] and references therein.

6.2 Calculating LRs

6.2.1 The match probability

Assume that the evidence under consideration, E, consists only of the information that the DNA profile obtained from the crime sample is D and the alleged source of that sample, Q, also has profile D. The possible effects of other profiled individuals are discussed in Section 6.2.6. Introducing the notation $\mathcal{G}_X$ to denote the genotype of X, the evidence E can be written succinctly as crime scene profile (CSP) = $\mathcal{G}_Q = D$, and the LR (3.2) becomes

$$R_X = \frac{P(\text{CSP} = \mathcal{G}_Q = D \mid H_X, E_{\text{o}})}{P(\text{CSP} = \mathcal{G}_Q = D \mid H_Q, E_{\text{o}})}, \tag{6.1}$$

where E_{o} denotes the other evidence and background information (we are continuing to assume that the DNA evidence is assessed last).

Here, we initially ignore the possibility of error (until Section 6.3.4). Thus, under H_Q, CSP = $\mathcal{G}_Q = D$ is equivalent to just $\mathcal{G}_Q = D$, and under H_X, CSP = $\mathcal{G}_Q = D$ is equivalent to $\mathcal{G}_X = \mathcal{G}_Q = D$. Assuming also that the fact that an individual committed the crime does not of itself alter the probability that they have a particular profile, (6.1) can be simplified further to

$$R_X = \frac{P(\mathcal{G}_X = \mathcal{G}_Q = D \mid E_{\text{o}})}{P(\mathcal{G}_Q = D \mid E_{\text{o}})} \tag{6.2}$$

$$= P(\mathcal{G}_X = D \mid \mathcal{G}_Q = D, E_{\text{o}}). \tag{6.3}$$

Thus, the LR R_X reduces to the conditional probability, called the 'match probability', that X has the profile *given* that Q has it: population genetic effects arise and can be dealt with via this conditioning. Other evidence E_0 such as eyewitness reports and alibis are typically irrelevant to DNA profile match probabilities. However, E_0 could include:

- information about the relatedness of Q with some other individuals,

- allele frequency information from population databases of DNA profile, and

- other relevant population genetics data and theory;

and these are potentially important for match probabilities.

The important feature of the match probability is that it takes account of *both* the observed profiles that form the match. Some authors, including the authors of the NRC report (Section 11.5), misleadingly refer to the population proportion of the profile as a 'match probability' which is inappropriate since the concept of 'match' involves two profiles, rather than just one. Equation (6.3) indicates that the question relevant to forensic identification is not

> 'what is the probability of observing a particular profile?'

but

> 'given that I have observed this profile, what is the probability that another (unprofiled) individual will also have it?'.

To emphasise the distinction, Weir [1994] employed the term 'conditional genotype frequency'.

In this section, we answer this question using the statistical model and sampling formula introduced in Section 5.3. Recall that the sampling formula follows from the assumption that subpopulation allele proportions have the beta (or Dirichlet) distribution with mean p and variance $F_{ST}p(1 - p)$. The value of F_{ST} can be interpreted either in terms of the mean square error of subpopulation allele proportions (Section 5.3) or in terms of coancestry in a simple migration-drift or mutation-drift model (Section 5.6).

Although F_{ST} cannot encapsulate all important population genetics phenomena, it does capture the essentials relevant to forensic match probabilities. Selection acting jointly at two or more loci each linked with a short tandem repeat (STR) locus can distort population profile proportions away from estimates based on assumption of independence across loci. However, this is a priori unlikely, and there is no reason to expect any such effects to systematically favour or disfavour defendants. The only population genetics phenomenon that, if ignored, systematically disfavours defendants is coancestry between defendant and alternative possible culprits, and that is what F_{ST} is intended to account for. Human population genetics is complicated, and inevitably, F_{ST} is an imperfect measure, but by choosing a suitable value defendants will not be systematically disfavoured, while match probabilities remain small enough to form the basis of satisfactory prosecutions in most cases.

6.2.2   Single locus

Suppose that CSP $= \mathcal{G}_Q =$ AA. If we assume that X and Q are unrelated, both come from the same subpopulation and neither is inbred, then the numerator of (6.2) corresponds to the probability (5.10) that a sample of four alleles from the subpopulation is AAAA. Dividing by the probability (5.7) that a sample of two alleles is AA, we obtain

---

Single locus match probability: CSP $= \mathcal{G}_Q =$ AA

$$\frac{(2F_{ST} + (1 - F_{ST})p_A)(3F_{ST} + (1 - F_{ST})p_A)}{(1 + F_{ST})(1 + 2F_{ST})}. \tag{6.4}$$

In the heterozygous case, under the same assumptions, we need the probability that in a sample of size four, both the first and second pairs of alleles are AB, divided by the probability that a sample of size two is AB. These probabilities are given at (5.8) and (5.9), except that we have to multiply by 2 for each pair of alleles (because of the two possible allele orderings), which gives

---

Single locus match probability: CSP $= \mathcal{G}_Q =$ AB

$$2\frac{(F_{ST} + (1 - F_{ST})p_A)(F_{ST} + (1 - F_{ST})p_B)}{(1 + F_{ST})(1 + 2F_{ST})}. \tag{6.5}$$

These match probability formulas are conditional on the values of F_{ST} and the p_j. Thus, strictly, they only apply if these parameters are known exactly, which is never the case in practice. In a fully Bayesian approach, (6.4) and (6.5) should be integrated (i.e. averaged) with respect to probability distributions for the unknown values of F_{ST} and the p_j. These probability distributions should be based on the available background information about the parameters, for example, from forensic DNA profile databases as well as population genetics theory and data. We prefer a simpler and more interpretable approach in which this background information is used to obtain estimates for F_{ST} and the p_j, which are then 'plugged in' to the conditional formulas. Care is required to choose the most appropriate 'plug-in' values for the parameters, as these need not satisfy the usual criteria for statistical estimators (see Section 6.3).

Use of (6.4) and (6.5) implies an assumption of Hardy–Weinberg equilibrium (HWE) within subpopulations, but not in the broader population from which the forensic database is drawn. Thus, as we have noted, testing for deviation from HWE within the broader population is of little value, whereas testing within the subpopulation is not usually feasible. Ayres and Overall [1999] have developed match probability formulas that take into account inbreeding within subpopulations. These formulas are more complicated than (6.4) and (6.5) but, nevertheless, relatively easy to apply. In the presence of within-subpopulation inbreeding, the formulas of Ayres and Overall [1999] indicate that match probabilities are slightly increased for homozygotes and decreased for heterozygotes; the overall effect on profile match probabilities is usually small but may be worth taking into account when populations practising marriage between relatives are relevant to a case.

Some single-locus match probabilities based on (6.4) and (6.5) are shown in Figure 6.1 for F_{ST} ranging from 0 to 1. Notice that increasing F_{ST} does not necessarily increase the match probability in the heterozygous case, although this almost always occurs in practice. Whatever the values of p_1 and p_2, as F_{ST} approaches 1, the homozygous and heterozygous match probabilities approach 1 and 1/3, respectively. Table 6.1 gives match probabilities for a possible genotype at each of four STR loci and for several values of F_{ST}. Very roughly, the single-locus match probabilities with $F_{ST} = 5\%$ are about 50% higher than that when $F_{ST} = 0$.

6.2.3   Multiple loci: the 'product rule'

There has been much debate in the forensic science literature about the validity of the 'product rule' for combining DNA match probabilities across loci. Combining probabilities via multiplication implies an assumption of statistical independence, and so the debate is equivalent to a debate about the *independence* of STR genotypes at different loci. This question can be rephrased:

> For two distinct individuals X and Q, does the event that their genotypes match at one or more STR loci affect the probability that they will match at the next locus?

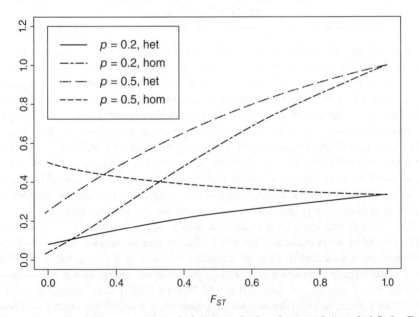

Figure 6.1   Single-locus match probabilities calculated using (6.4) and (6.5) for F_{ST} ranging from 0 to 1. In the heterozygote case, the proportions of the two alleles are both equal to p.

Table 6.1   Single-locus match probabilities (after multiplication by 10^3) assuming X unrelated to Q, for four STR loci and various values of F_{ST}, for the alleles specified in column 2 and assuming the population proportions given in column 3.

STR locus	Geno-type	Population proportions	Match probability $\times 10^3$			
			$F_{ST} = 0$	$F_{ST} = 1\%$	$F_{ST} = 2\%$	$F_{ST} = 5\%$
D18	14, 16	0.16, 0.14	45	49	52	64
D21	28, 31	0.23, 0.07	32	37	41	54
THO1	9.3, 9.3	0.30, 0.30	90	101	112	145
D8	10, 13	0.09, 0.33	59	65	70	85

It is important to appreciate that 'independence' is not an absolute concept but can depend on the modelling assumptions employed in deriving the probabilities. In children of mixed ages, reading ability and shoe size are not independent, because older children have, on average, both larger shoe sizes and higher reading abilities. However, in a statistical analysis that adjusts for age, we expect shoe size and reading ability to be (at least approximately) independent, because we believe that there is no causal connection between them.

For STR matches, the relevant extrinsic factor, the equivalent of the child's age in the aforementioned example, is the degree of relatedness between the two individuals. If X and Q are directly related through one or more known common ancestors (e.g. grandparents), then matches at distinct loci are not independent, but an appropriate adjustment for the relationship can restore approximate independence (see Section 6.2.4). All humans are related at some level, and if X and Q are apparently unrelated, this only means that the relationship between them is unknown and presumed to be distant. However, it is not necessarily distant enough to be negligible for the purposes of calculating match probabilities. You and I could well have a common ancestor as little as five generations ago: few people fully know their true ancestry that far back, and thanks to international wars, colonisations and migrations, this possibility remains even if we were born in different parts of the world.

The question

'Is the product rule valid[1] for forensic STR loci?'

does not have a clear answer without making the question more precise. If the product rule is understood to mean multiplying together match probabilities based on an assumption of no relatedness, then the answer is a clear 'no'. However, the match probabilities (6.4) and (6.5) are conditional on a level of coancestry between apparently unrelated individuals measured by F_{ST}. we have argued in Chapter 5 that coancestry is the only population genetic factor that causes a deviation away

---

[1] Here, 'valid' is understood to mean 'provides a satisfactory approximation'. In particular, 'satisfactory' implies 'not systematically biased against defendants'; some bias in favour of defendants is usually regarded as acceptable, see the discussion in Section 6.3

from independence systematically disfavouring defendants, and it follows that the product rule is valid given these match probability formulas and an appropriate value for F_{ST}. Use of the product rule with an F_{ST} adjustment implies an assumption of linkage equilibrium (LD) (Section 5.5) within subpopulations but not in the broader population. The latter may manifest LD due to subdivision and drift, or admixture, that can be accounted for via F_{ST}.

Although match probabilities at many loci cannot readily be checked, the available population genetics theory and data support the view that given a suitable value of F_{ST}, match probabilities obtained as products of (6.4) and (6.5) will not be systematically unfavourable to defendants. If this argument is accepted, then match probabilities at multiple loci can be obtained by multiplying together the match probabilities at the individual loci calculated using (6.4) and (6.5).

Ayres [2000a] investigated the effects of inbreeding within subpopulations on two-locus match probabilities. She found that multiplying F_{ST} adjusted single-locus match probabilities is conservative for double heterozygotes but not for double homozygotes. The overall effect is small relative to the effect of using an F_{ST} adjustment, particularly when a conservative value is used for F_{ST}. Here, and subsequently, 'conservative' applies to approximations or estimates that tend to err in favour of the defendant; for a discussion, see Section 6.3.

Whole-profile match probabilities using the F_{ST}-adjusted product rule applied to the four STR loci of Table 6.1, and using various values of F_{ST}, are given in the final row of Table 6.2. Assuming $F_{ST} = 5\%$ increases the four-locus match probability over fivefold relative to the $F_{ST} = 0$ case, consistent with approximately 50% in the single-locus values. This extrapolates to more than a 50-fold increase for 10 loci, and over 400-fold for a 15-locus profile match.

6.2.4 Relatives of Q

So far, we have been considering alternative possible culprits X that are unrelated to Q. Here we consider the possibility that X and Q are closely related. We continue to assume that the DNA profile of Q is available to the court but that of X is not: it would

Table 6.2 Match probabilities for the four-locus STR profile of Table 6.1 under some possible relationships of Q and X and for various values of F_{ST}.

Relationship	Match probability			
	$F_{ST} = 0$	$F_{ST} = 1\%$	$F_{ST} = 2\%$	$F_{ST} = 5\%$
Identical twin	1	1	1	1
Sibling	17	19	20	23×10^{-3}
Parent/child	14	17	19	29×10^{-4}
Half-sib	23	29	35	61×10^{-5}
Cousin	6	8	10	20×10^{-5}
Unrelated	8	12	17	43×10^{-6}

in principle be desirable to exclude close relatives of Q from suspicion by examining their DNA profiles, but this is rarely possible in practice.

Let Z denote the number of alleles at a locus that X and Q share directly from a known, recent common ancestor (e.g. parent or grandparent) and let

$$\kappa_j = P(Z = j),$$

for $j = 0, 1, 2$. The probability distribution for Z under some common, regular relationships (i.e. no inbreeding) are shown in Table 6.3.

If $Z = 0$, we are in the situation described in Section 6.2.2 (Q and X unrelated) and we write M_2 for the appropriate match probability, either (6.4) or (6.5). If $Z = 2$, a match is certain. For $Z = 1$, consider first the case of an AA homozygote match. Although there is a total of four alleles at the locus, two of these are shared from the recent common ancestor, and so effectively we have observed three matching alleles. The match probability is the probability of this observation given that the first two alleles match, which using (5.6) is

$$M_1 = \frac{2F_{ST} + (1 - F_{ST})p_A}{1 + F_{ST}}.$$

In the heterozygous case, there are two equally likely possibilities (A or B) for the allele that is shared from the recent common ancestor, and the match probability is the probability that the third allele sampled is the other one of A and B. Thus

$$M_1 = \frac{F_{ST} + (1 - F_{ST})(p_A + p_B)/2}{1 + F_{ST}}.$$

The overall single-locus match probability for relatives is then

$$P(Z = 2) + M_1 P(Z = 1) + M_2 P(Z = 0), \tag{6.6}$$

and whole-profile match probabilities can be obtained by multiplication over loci, provided that the loci are unlinked. See Section 7.1.4 for a discussion of the effects of linkage.

Table 6.3   Relatedness coefficients for some regular relatives.

Relationship	κ_0	κ_1	κ_2	$\bar{\kappa}$
Sibling	1/4	1/2	1/4	1/2
Parent	0	1	0	1/2
Half-sib	1/2	1/2	0	1/4
Cousin	3/4	1/4	0	1/8
Unrelated	1	0	0	0

$\bar{\kappa}$ is half the expected number of alleles shared identical by descent (IBD) between the relatives, which is twice Malécot's 'coefficient of kinship'. The coefficients for aunt/uncle–niece/nephew and grandparent–grandchild are the same as those for half-sibs.

Four-locus match probabilities based on (6.6) and multiplication over loci are shown in Table 6.2 for the four STR loci of Table 6.1. Notice that the value of F_{ST} still affects the match probability even when X and Q are assumed to be closely related. This is because, as well as matches arising via alleles shared from the recent known ancestors of X and Q, alleles can also be shared from more distant common ancestors. The latter possibility is accounted for in the F_{ST} value. The relative importance of F_{ST} declines as the known relationship between X and Q becomes closer. See Balding and Nichols [1994] for further discussion.

6.2.5   Confidence limits †

Some authors have discussed the need for confidence limits, or similar measures of uncertainty, on LRs. Forensic scientists will be aware that confidence intervals are routinely used to measure the uncertainty about an unknown parameter in many areas of science. However, note an important distinction: if you are interested in an unknown quantity that has a continuous range of possible values – say the distance to the moon – then it makes sense to give an estimate of the uncertainty in any particular value. In the Bayesian approach, every possible value for the distance to the moon is assigned a probability (density), and these can be used to compute, say, a shortest 95% probability interval for the unknown value. The probability assignments represent uncertainty about the true distance, and there is little benefit in measuring your uncertainty about this uncertainty.

An LR R_X is primarily concerned with an unknown that has only two possible values: is Q or X the source of the crime scene DNA? The probabilities used to formulate the LR represent uncertainty about the correct answer, and there is little benefit from trying to measure the uncertainty about this uncertainty.

The relative frequency of an allele or entire profile, either the actual population proportion or the theoretical value under an evolutionary model, is an unknown for which it does make sense to give some measure of uncertainty about an estimated value, for example, a confidence interval. The rationale behind confidence intervals is similar to that underpinning classical hypothesis testing, to be discussed in Section 10.3. It is based on imagining a long sequence of similar 'experiments'. Roughly speaking, a 95% confidence interval is calculated via a rule that has the property that in 95% of repeated 'experiments', the computed confidence interval would include the true value.

If relatedness and F_{ST} are ignored, we can imagine an experiment of choosing alleles at random in the population, and calculate a confidence interval for a profile proportion that takes only the uncertainty due to sampling into account. This may be achieved, for example, using the asymptotic normality of its logarithm [Chakraborty et al., 1993]. However, as noted by Curran et al. [2002], the approach has severe limitations in more complex scenarios, involving, for example, relatedness and mixtures. In such settings, it can be difficult to specify the imaginary sequence of similar cases.

We have argued in Section 6.2.1 that although estimates of population allele or profile proportions can be helpful in formulating the required probabilities, they are not in themselves the directly relevant quantities for assessing the weight of evidence. It follows that confidence intervals for them are of little use. Buckleton et al. [2005] suggested that irrespective of the philosophical arguments, a forensic scientist is likely to be asked for a confidence interval, or some similar measure of uncertainty about a reported LR, and he/she may appear unscientific if he/she fails to provide it. This does not correspond to our experience. There is possibly no harm in providing a confidence interval for a population profile proportion if a court calls for it, but a practice of routinely reporting such a confidence interval offends against the principle that a forensic scientist should avoid, as far as possible, presenting unnecessary information.

There is a legitimate concern that any calculation of an LR requires assumptions and data, and two equally competent forensic scientists may employ (slightly) different assumptions and data and, hence, arrive at different LRs. A court may wish to have an indication of alternative reasonable assumptions and how big an effect they may have on the LR. What matters most for forensic match probabilities is the level of coancestry between X and Q, and uncertainty about any specific profile match probability can best be conveyed by exploring different assumptions about the value of F_{ST}. We believe that giving a range of likelihood values under different assumptions about F_{ST} satisfactorily addresses the issue of uncertainty about a particular value for the LR.

6.2.6 Other profiled individuals

So far we have assumed that the DNA evidence consists only of the profile of Q and the CSP. In practice, however, the profiles of many other individuals are potentially relevant to the match probability.

Observed frequencies of STR alleles in population databases of anonymous individuals form an important source of information relevant to match probabilities. In the approach based on 'plug-in' estimates for F_{ST} and the p_j (see page 85), these data enter the match probability only via the parameter estimates, to be discussed in Section 6.3. We noted when introducing the plug-in approach that integrating the match probability with respect to a probability distribution for the parameters is preferable, at least in principle, though more complex. Pueschel [2001] attempts a full probability approach, explicitly conditioning the match probability on all the DNA profiles in a database. He takes population subdivision into account but does not give any special status to the profile of Q, neglecting the fact that this profile has a particular relevance to the crime that is not shared by the database profiles.

In addition to population databases of anonymous profiles, the profiles of known individuals may have been considered during the crime investigation. We assume that none of these profiles matched that of Q (and hence also the CSP), otherwise no case can proceed against Q without substantial further evidence incriminating him and not the other matching individual.

The excluded individuals may have been possible suspects profiled in the course of the crime investigation, or their profiles may have already been recorded in an intelligence database. In either case, as discussed in Section 3.4.5, the effect of every excluded possible suspect is to (slightly) increase the overall case against the defendant Q, since an alternative to the hypothesis that Q is guilty is thereby eliminated, removing a term from the denominator of (3.3). In addition, the observation of non-matching profiles slightly strengthens the belief that the observed CSP is rare. Both these effects are typically very small and it would usually seem preferable to neglect the observed, non-matching, profiles when calculating match probabilities. This practice is slightly favourable to defendants and, therefore, need not be reported to the court, nor any adjustment made to the LR because of it.

6.3   Application to STR profiles

According to the weight-of-evidence formula (3.3), a rational juror should assess an LR for every alternative possible culprit X. Under the assumptions of Section 6.2, these LRs are equivalent to match probabilities, and we have derived formulas for them in terms of population allele proportions (the p_j) and F_{ST}. Instead of integrating the match probability over these unknown values, with respect to a probability distribution representing uncertainty about their values for the alternative suspects in a particular case, we advocate here a simpler approximation obtained by 'plugging in' estimates for these parameters. We now consider the choice of plug-in values.

One possible reason for a forensic scientist to use relatively large values for F_{ST} and the p_j is a desire on the part of courts to be conservative (tend to favour the defendant). Excessive conservatism could lead to an unnecessary and undesirable failure to convict the guilty, and it is difficult to formulate principles for how much conservatism is appropriate. Despite this difficulty, it does seem worthwhile, at least for criminal cases, to adhere to the principle that the prosecution should not, as far as reasonably possible, overstate its case against a defendant.

6.3.1   Values for the p_j

In typical forensic identification problems, we have available estimates of allele proportions in broadly defined populations, such as Caucasians, sub-Saharan Africans, East Asians or Aboriginals. Recall that in the single-contributor identification setting, the match probability (6.3) is the conditional probability that X has profile D, given that Q has it. It follows that, in principle, the population database to be used to estimate the p_j should be the most appropriate for X, the alternative possible culprit under consideration. It is simpler, and usually conservative, to use the database closest to the suspect Q for all X, together with an appropriate value for F_{ST} [Steele and Balding, 2014b]. There is sometimes difficulty in choosing the database most appropriate for Q, for example, because he/she is of mixed race or from an admixed population, or a minority group, or simply because the population from which the database is drawn is not well defined. If so, it may be acceptable to use the database yielding the largest match probability. The additional uncertainty arising in such cases can be allowed for in choosing the value of F_{ST}: the 'worse' the database in reflecting the genetic background of a possible culprit, the greater is the appropriate value of F_{ST}.

Forensic population databases usually consist of at least several hundred profiles. This is large enough that allele proportions above 1% can be estimated reasonably well, provided that the sample is representative (see Section 6.3.2). However, there remains some sampling uncertainty, whose effects on match probabilities need to be accounted for (Section 2.3.1).

In applying (6.4) and (6.5) to actual cases, it seems natural to estimate p_j, the population proportion of allele j, by its relative frequency in the database most relevant to the alternative possible culprit X. The database closest to Q will, in general, provide a conservative alternative for all X, and the difference should be small given an appropriate value for F_{ST} (see below).

The sampling uncertainty in the estimate of p_j still needs to be taken into account (Section 2.3.1). One approach to this (see Balding and Nichols, 1994) is to estimate p at a heterozygous locus by

$$\widehat{p_j} = \frac{x_j + 2}{n + 4},$$

(6.7)

where x_j is the frequency of allele j in the database and n is the total frequency of all alleles: $n = \sum_j x_j$ (see Balding and Nichols, 1994, 1995). In the homozygous case, an analogous estimate is

$$\widehat{p_j} = \frac{x_j + 4}{n + 4}.$$

(6.8)

They can be justified as approximations to the posterior mean given the sample allele counts and a (multivariate) uniform prior distribution[2] for the allele proportions. The effects of using these estimates, rather than the database proportions x_j/n, are

(i) smoothing out some sampling variation for low-frequency alleles, and

(ii) building in some conservatism by increasing the estimates of the p_j for the observed alleles by a relatively large amount when they are rare.

As a direct consequence of (ii), the allele proportions used in different cases sum to more than 1. This property also applies to the use of upper confidence limits, or any other conservative estimators, in place of the p_j.

Curran et al. [2002] are critical of the approach based on using the 'plug-in' estimates (6.7) and (6.8) in place of a more careful accounting for the sampling uncertainty in estimates of the p_j. We acknowledge that it is a rough rule-of-thumb approach. It is neither intended to serve as an approximate confidence limit nor indeed to satisfy any specific criterion of rationality. We believe that Curran *et al.* will agree with us that the 'gold standard' is to integrate over a full probability distribution for the p_j, given the database frequencies and a reasonable prior distribution. However, this can be computationally intensive for little gain, and there are interpretational advantages in using explicit values for the p_j. Moreover, uncertainty about the p_j is in practice relatively unimportant compared with the effects of the value of F_{ST}.

---

[2] More generally, the Dirichlet distribution.

6.3.2   The value of F_{ST}

In Section 5.2, we defined F_{ST} in terms of the mean square error of the allele proportion in a subpopulation to which X and Q both belong, relative to a reference value such as the allele proportion in the population from which the available database was drawn. In Section 5.6, we showed that the value for F_{ST} can also be interpreted in terms of the coancestry of X with Q, and in Section 5.7, we showed that estimates of F_{ST} can depend sensitively on the choice of reference values.

Published estimates of F_{ST} at STR loci for subpopulations relative to a continental scale population are often small and typically less than 1%. Nevertheless, there are several arguments for using larger values in forensic practice:

- The appropriate value of F_{ST} is never available, and we are reduced to using an educated guess. Among the sources of uncertainty is that subpopulations are never well defined and nor the populations from which the reference databases are obtained. In order to avoid unfairness to the defendant, we should prefer to over-estimate rather than under-estimate F_{ST}.

- Strictly, we should integrate the match probability over a distribution of values for F_{ST}, and because the integration is a product over many terms, the integral will be dominated by the upper tail of the distribution.

- Many published estimates of F_{ST} usually relate to the variation of allele fractions around the observed mean value. However, in forensic applications, what is needed is the variation of subpopulation values away from the forensic database value, and this may well be substantially larger. In particular, minority ethnic groups may be heterogeneous and the database allele frequency may not be representative of the specific ethnic group relevant to a particular crime.

- Although we have introduced F_{ST} in terms of the variance of subpopulation allele fractions, subpopulations are a theoretical construct and difficult to specify precisely. (In the United Kingdom, do Roman Catholics form a subpopulation? do the people of Surrey form a subpopulation?). The coancestry interpretation (Section 5.6) is usually more useful, in which case F_{ST} pertains to the individual X rather than his/her subpopulation. The individuals with larger values of F_{ST} will contribute disproportionately to the total weight of evidence in formula (3.3), and this can be allowed for by using a 'generous' value of F_{ST}.

Steele et al. [2014b] used a dataset of worldwide STR genotypes to estimate F_{ST} values for a range of subpopulations at approximately the national scale relative to continental-scale populations. They found that $F_{ST} = 1\%$ is sufficient for European Caucasian subpopulations, but higher values are appropriate for European diaspora subpopulations (such as the United States or Latin America). $F_{ST} = 3\%$ suffices for almost all subpopulations relative to the most appropriate populations. A larger value, perhaps 5%, may be necessary for isolated minority groups such as Sardinians. The work of Steele et al. raises the possibility of tailoring the F_{ST} value used to that most appropriate for a given nationality, for example, the F_{ST} value for a Kenyan Q with an African database may need to be greater than that for a Nigerian Q.

If X is not from the same population as Q, then they have little coancestry and so a small value of F_{ST} can be justified. However, since the problem of representativeness of any database remains, as well as possibilities for some coancestry even across populations, we advocate using a non-zero value of F_{ST} in every case, perhaps setting 1% as the minimum.

6.3.3   Choice of population

Typically, there is little information to guide the choice of population allele fre-quency databases for an unknown contributor. In the United Kingdom, five such databases are routinely available (see Figure 5.2). One approach is to evaluate the LR with all possible databases in turn and to report the result that is most favourable to the defence. This approach becomes impractical with a large number of databases and for complex profiles involving multiple unknown contributors. Moreover, the database that produces the most favourable LR may be incongruent with the circumstances of the case. For instance, if a crime was committed in a rural community in the United Kingdom with predominantly Caucasian ancestry and Q is from that community, does it make sense to employ an East Asian database for X because it provides the most favourable LR for the defence? We propose instead for all unknown contributors the use of the database most appropriate of Q, together with a moderately large value of F_{ST} (we suggest that $F_{ST} = 0.03$ is often appropriate [Steele and Balding, 2014b]). We show in a simulation study that this heuristic is conservative compared to assuming the true database for each contributor (which may or may not be the database most appropriate for Q).

6.3.4   Errors

We saw in Section 4.1.1 that there are a number of factors that can lead to false exclusion error due to a culprit's STR profile being reported incorrectly from either the crime or the evidence sample. The overall false exclusion rate is thought to be extremely small and is negligible compared with the rate of unsolved crimes. In any case, we saw in Section 2.3.3 that having observed a match, the possibility of a false mismatch is almost irrelevant to evidential weight.

Potentially much more important are false inclusion errors, in which an innocent individual is reported to have a profile that does not exclude him from being the culprit. As we noted in Section 3.4.4, there are at least two ways in which this can occur: a chance match, the probability of which is assessed by the match probabilities calculated in Section 6.2, or a false match, perhaps due to an incorrect recording of profiles from either crime or evidence samples. We also noted in Section 3.4.4 that

(i) the match probability is effectively irrelevant if swamped by the probability of a false-match error;

(ii) it is not the probability of any error that matters but only an error that leads to a false inclusion.

Some critics of the reporting of DNA profile matches have argued that (i) above is routinely the case, and hence the match probabilities reported in court are irrelevant and potentially misleading. This argument has some force, and protagonists of DNA profiling have been prone to mistakes and exaggeration in their attempts to discount it (e.g. see Koehler, 1996). However, because of (ii), the chance of a false inclusion error due to genotyping anomalies is so remote as to be negligible, even relative to the match probability. Contamination is a distinct possibility in some settings, but because evidence and crime samples are routinely typed by different staff in different laboratories, sometimes with a substantial time gap, in many cases, this possibility can also reasonably be ruled out.

This leaves the possibility of false inclusion due to handling or labelling error, or evidence tampering, 'frame-ups', fraud and similar. A conspiracy theory, such as that police and/or judicial authorities colluded to manufacture evidence falsely linking the defendant with the crime scene, may be regarded by a reasonable juror as substantially more plausible than a chance match whose probability is reported as less than 1 in a billion, even when there are no particular reasons to suggest a conspiracy.[3] The LR for the DNA evidence under such a conspiracy theory is likely to be close to 1, so that the juror may as well dismiss the match probability as irrelevant (see Exercise 2, Section 6.6). The relevant probability for the juror to assess is the probability of such a conspiracy based on all the other evidence.

There seems no role for a forensic scientist to predict whether or not a juror might pursue such a line of reasoning, and hence, the only reasonable option is to supply the juror with a match probability and also to try convey an understanding of the circumstances under which the match probability would be effectively irrelevant.

6.4   Application to haploid profiles

6.4.1   mtDNA profiles

Profiles based on mitochondrial DNA (mtDNA) sequences were introduced in Section 4.2. Because of the absence of recombination, the whole mtDNA sequence can be considered a single locus with many possible alleles. Unlike autosomal loci, at which we each carry two alleles, we normally carry just one mtDNA allele.

The uniparental inheritance of mtDNA has advantages in establishing the related-ness of individuals through female lines: close maternal relatives of an individual will have (almost) the same mtDNA. This feature was used to help to identify the remains of the Russian royal family [Gill et al., 1994], in part using mtDNA from a member of the British royal family. However, this feature can be a disadvantage for forensic identification, because an individual cannot easily be distinguished from any of his/her maternal relatives, even those removed by tens of generations. Moreover, a person's maternal relatives are likely to be unevenly distributed geographically, and

---

[3] The case of OJ Simpson (California, 1995) provides an example in which the defence argued that there was evidence of a conspiracy, and the jury acquitted. See the discussion in Buckleton et al. [2005].

it will often be difficult to assess how many close maternal relatives of Q should be considered as alternative possible sources of an evidence sample. The relatively high mutation rate of many mtDNA sites lessens this problem but does not eliminate it. On the other hand, the high mutation rate leads to the problem of heteroplasmy (multiple mtDNA types in the same individual). As noted in Section 4.2, heteroplasmy can be problematic for inference because not all of an individual's mtDNA sequences may be observed in, for example, typing a small and/or degraded CSP.

Currently, the most widespread approach to interpret mtDNA profiles in an identification problem in which there is a match between suspect and crime-scene mitotypes ($\equiv$ mtDNA types) is to report this evidence along with the frequency of the mitotype in a population database of size N. This simple approach fails to deal adequately with the complications outlined above, as we now briefly discuss.

To make some allowance for sampling variability, it is advantageous to include both the CSP and the defendant profile with those of the population database. For autosomal markers, this practice leads to the 'pseudo-count' estimates (6.7) and (6.8). Here, we obtain

$$\hat{p} = \frac{x+2}{N+2}, \tag{6.9}$$

where x denotes the database count of the mitotype. Wilson et al. [2003] considered a sophisticated genealogical model that could exploit the sample frequencies of similar profiles to refine estimates of p, but they found that $\hat{p}$ given at (6.9) provided a satisfactory, simpler estimator that is typically slightly conservative.

The above analyses take no account of the effects of any population substructure. Using an approach analogous to that used to derive (6.4) and (6.5), the match probability for an mtDNA allele with population frequency p is obtained by dividing the case $m = n = 2$ of the sampling formula (5.6) by the case $m = n = 1$, which gives

---

Match probability: haploid case

$$m_X = F_{ST} + (1 - F_{ST})p. \tag{6.10}$$

Since p is typically unknown, (6.10) can be combined with (6.9) to obtain a practical formula. This gives essentially the formula reported as an 'intuitive guess' by Buckleton et al. [2005].

Formula (6.10) applies to a structured population in which the subpopulations are well mixed. It does not take account of the fact that maternally related individuals might be expected to be tightly clustered, possibly on a fine geographical scale. Reports of F_{ST} estimates for mtDNA drawn from cosmopolitan European populations typically cite low values, reflecting the fact that this population is reasonably well mixed, as well as the effects of high mtDNA mutation rates. However, researchers are rarely able to focus on the fine geographic scale that may be relevant in forensic work, and there are some large F_{ST} estimates at this scale. See Buckleton et al. [2005] for a discussion on mtDNA F_{ST} estimates and their limitations.

In the presence of a matching heteroplasmic profile, no modification to (6.10) is required except that p refers to the population proportion of individuals displaying the same heteroplasmy. The value of p will thus be difficult to estimate and will in part depend on the typing platform employed. Imperfect matches can arise with heteroplasmic profiles. For example, the crime scene mtDNA profile may consist of just one sequence, say d, whereas the mtDNA profile of Q displays a set of sequences D that includes d. In this case, the denominator of the LR (3.2) requires an assessment of the probability that under the circumstances of the recovery of crime scene DNA, only d would be recorded rather than the full, heteroplasmic, profile D, if Q is the culprit. The numerator of the LR (X is the culprit) requires an estimate of the joint probability of the two observed mitotypes given that X and Q are distinct (but possibly related) individuals.

Although selection is thought likely to influence the distribution of mtDNA types, because no assumption of either Hardy–Weinberg or linkage equilibrium is made in calculating mtDNA match probabilities, it seems reasonable to assume that selection will not adversely affect the validity of (6.10).

6.4.2   Y-chromosome markers

See Section 4.3 for a brief introduction to the properties of Y-chromosome markers. The interpretation issues for Y-chromosome profiles are similar to those for mtDNA.

The problem of heteroplasmy rarely, if ever, arises for Y chromosomes, but the problem of population structure is even more important [Jobling et al., 1997] since values of F_{ST} are typically higher than that for mtDNA. Zerjal et al. [2003] reported a specific Y haplotype that is frequent in many parts of Asia, and largely unobserved elsewhere, and which they suggested can be attributed to the consequences of historic Mongol conquests, possibly even to Genghis Khan himself. Similar historical events on a smaller scale may have led to an unrecognised, local elevation of the concentration of a specific Y haplotype that is otherwise rare. Microgeographical variation in forensic Y-chromosome haplotypes has been studied by Zarrabeitia et al. [2003] in the Cantabria region of Spain. These authors found that substantial overstatement of evidential strength frequently results from the use of population databases collected on too broad a geographical scale.

Brenner [2014] discussed many issues relevant to the interpretation of Y-STR profiles, including an approach based on coalescent theory [Andersen et al., 2013], but these authors do not consider the geographical heterogeneity problem that we highlight here.

6.5   Mixtures

6.5.1   Visual interpretation of mixed profiles

A mixed profile arises when two or more individuals contribute DNA to a sample. An example of an electropherogram (epg) corresponding to a mixed STR profile is shown in Figure 6.2. The profile at the amelogenin sex-distinguishing locus (leftmost on Figure 6.2b) shows a predominance of X, but some trace of Y, suggesting that the mixture stain may have come predominantly from a female, with a minor contribution from a male. This mixture was created in the laboratory, with known contributors, and the above interpretation is indeed correct, but let us proceed for the moment as if we did not know this.

The recorded signals at other loci are consistent with there being two contributors, one 'major' and the other 'minor'. For example, at locus VWA (second locus from the left in Figure 6.2a), the epg indicates the presence of four alleles, two of which (labelled 15 and 18) produce a strong signal, while the remaining two signals (corresponding to alleles 14 and 19) are much weaker, though still clearly distinguishable from the background noise.

Two other loci show similar four-allele patterns in the epg. However, at most of the loci, only two or three alleles appear in the epg, which can arise if one or both contributors is homozygous, or if they have alleles in common. At locus D2S1338 (rightmost in Figure 6.2a), there is a strong signal at allele 17, a slightly weaker signal at allele 19 and a much weaker signal at allele 18, suggesting that the major contributor has genotype 17,19 at this locus, while the minor contributor is 17,18. An alternative possibility is that the peak at allele 18 corresponds to a stutter peak, and only alleles 17 and 19 are actually present in the sample, but this is unlikely because the peak, although low, is higher than normal for a stutter peak at this position on the epg. Although the 17,19 and 17,18 seem the most plausible genotype designations for major and minor, respectively, it is difficult to assign any measure of confidence to this call.

Interpretation at the TH01 locus (third from left in Figure 6.2c) is even more difficult: the two observed alleles, 6 and 9.3, display allele signals of noticeably different heights, yet they are not extremely asymmetric. Perhaps the major contributor is 6,9.2 and the minor is a 6,6 homozygote, but other possibilities seem to exist, such as that both contributors are 6,9.3 and the apparent peak imbalance results from a fluctuation in the experimental conditions: here the ratio of peak heights is approximately 2/3, which may be regarded as within the normal range of variation of peak heights for a single, heterozygous contributor.

As the above discussion suggests, inferring the profiles of major and minor contributors to a sample can sometimes be done with reasonable confidence, but often it is problematic at least for some loci. The presence of good-quality DNA and a strong imbalance in the proportions of the DNA from each source individual facilitate the task. However, in the presence of degraded samples, low DNA copy number, an unknown number of contributors or an equal contribution from two contributors, the task can be challenging, and assigning a measure of confidence to any particular genotype designation is problematic.

6.5.2   Likelihood ratios under qualitative interpretation

One approach to overcoming the problems with visual interpretation of mixtures, at the cost of discarding the quantitative information from the epg, for example, about peak heights and shapes, is to limit interpretation to qualitative allele calling only, without any attempt to infer the underlying genotypes. Thus, the interpretation of

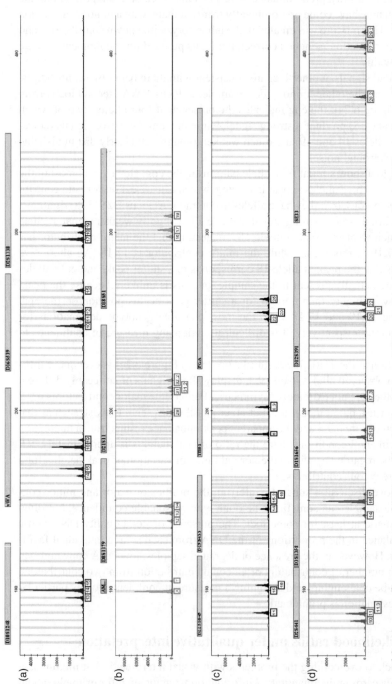

Figure 6.2   An epg showing a mixed STR profile obtained from a DNA sample with two contributors. The major profile is from a female and the minor component from a male. The dye colours represented in each panel are the same as for Figure 4.1, namely (from top to bottom): (a) blue, (b) green, (c) black and (d) red. The DNA sample was amplified with the NGM SElect®STR kit. The amplified fragments were separated on the ABI 3130xL Genetic Analyzer and analysed using the GeneMapper®3.2 software. Image supplied courtesy of Cellmark Forensics. ©2014 Cellmark Forensics.

locus VWA in the epg of Figure 6.2 would be limited to the conclusion that alleles 14, 15 and 16 are observed in the mixture. Then, all combinations of underlying genotypes that include at least one copy of each of these alleles are regarded as equally plausible.

We consider here the single-locus case; LRs can be combined across loci via multiplication (see Section 6.2.3). These approaches were initially developed by Evett et al. [1991] for non-STR profiles, but they remain applicable to STR profiles. Mortera et al. [2003] described a probabilistic expert system (Section 7.2.3) for the qualitative analysis of DNA mixtures, and the qualitative approach has been extended to take account of the coancestry of all the contributors to a mixture [Curran et al., 1999; Fung and Hu, 2000, 2002] and alternative suspects related to the accused [Fung and Hu, 2004]. These authors also offer software for mixture interpretation [Hu and Fung, 2003].

6.5.2.1   The number of unknown contributors

For STR profiles, a minimum number of contributors to a sample of DNA are provided by half the number of distinct alleles observed at any one locus. Even in the case that no more than two alleles are observed at any locus, it is possible that there is more than one contributor, although this is typically very unlikely. It is not possible to put an upper bound on the number of contributors, although it may be possible to estimate the number of contributors [Haned et al., 2011, Perez et al., 2011].

In principle, prior to the DNA evidence, judgements about the number of possible sources should lie in the domain of the court, not the DNA expert. However, there will inevitably be occasions when experts make such prior judgements, when they seem uncontroversial and when the alternative may be to overly complicate their evidence (e.g. by working out LRs under many different scenarios, most of which are implausible).

We distinguish three classes of contributors to a mixture:

- the contributor of interest, or 'contested contributor', who is either Q or X (we will write Q/X),

- known contributors, who will be denoted as $K, K1, K2, \ldots$, and

- unknown contributors denoted as $U, U1, U2, \ldots$.

Although 'known' and 'unknown' provide a convenient shorthand, the key point is not whether the person is known but whether their reference DNA profile is available for the evaluation. In some mixture CSPs, there may be multiple contested contributors, but we propose that those cases be tackled by computing a sequence of LRs each involving only one contested contributor.

6.5.2.2   Two contributors: Q/X and K

The easiest case arises when all the contributors to a mixture other than the contested contributor are known (DNA profiles available). Here we consider the common scenario of one known contributor (K) in addition to Q/X. Then the evidence is the CSP and the reference profiles of Q and K, denoted as G_Q and G_K. It is reasonable to assume that $\mathcal{G}_Q$ and $\mathcal{G}_K$ are equally likely under both hypotheses, and so the LR takes the form

$$R_X = \frac{P(CSP|\mathcal{G}_Q, \mathcal{G}_K; \ X \text{ and } K \text{ are the sources of the DNA})}{P(CSP|\mathcal{G}_Q, \mathcal{G}_K; \ Q \text{ and } K \text{ are the sources of the DNA})}. \tag{6.11}$$

At some loci, one or both alleles of X may be masked by the alleles of K. In general, the numerator of R_X is computed by summing over the possible genotypes of X, given the CSP and $\mathcal{G}_K$. Each term in the sum involves the conditional probability of a particular value of $\mathcal{G}_X$, given $\mathcal{G}_Q$ and $\mathcal{G}_K$ (the conditioning is irrelevant if $F_{ST} = 0$).

Example

Consider a single locus, and suppose that the following are observed:

$$\mathcal{G}_Q = AB \qquad \mathcal{G}_K = AC \qquad CSP = ABC.$$

Then (6.11) becomes

$$R_X = \frac{P(ABC \mid \mathcal{G}_Q = AB, \mathcal{G}_K = AC; \text{ sources: } X, K)}{P(ABC \mid \mathcal{G}_Q = AB, \mathcal{G}_K = AC; \text{ sources: } Q, K)}. \tag{6.12}$$

The denominator equals 1. For the numerator, given that $\mathcal{G}_K = AC$, the CSP implies that $\mathcal{G}_X$:

- includes a B allele, and

- does not include any allele other than A, B and C.

The possible genotypes consistent with these two requirements are AB, BB and BC. Thus, the numerator of (6.12) is the probability that X has one of these three genotypes, given that $\mathcal{G}_Q = AB$ and $\mathcal{G}_K = AC$. Here, we assume

- that Q, K and X are mutually unrelated;

- no coancestry (i.e. $F_{ST} = 0$);

- that the allele probabilities p_A, p_B and p_C are known; and

- genotypes are in Hardy–Weinberg proportions (see Section 5.4).

Then (6.12) becomes

$$R_X = P(\mathcal{G}_X = AB) + P(\mathcal{G}_X = BB) + P(\mathcal{G}_X = BC)$$
$$= 2p_A p_B + p_B^2 + 2p_B p_C = p_B(2p_A + p_B + 2p_C).$$

Since $F_{ST} = 0$, possible coancestry between X and one or both of Q and K is ignored. If X, Q and K are all assumed to have a common level of coancestry measured by F_{ST}, then we need to take into account the four alleles (AABC) already observed in Q and K, so that

$$R_X = 2P(AB \mid AABC) + P(BB \mid AABC) + 2P(BC \mid AABC), \tag{6.13}$$

where each probability is for an ordered sequence of two alleles. Just as for the derivations of (6.4) and (6.5), the conditional probabilities of (6.13) can be evaluated using the sampling formula, (5.6). For example, P(AB | AABC) is the probability of observing A then B in two further draws from a population, when a sample of size four has already been observed to be AABC. This probability can be computed as the product of instances of (5.6) with $m = 2$ and $n = 4$, and with $m = 1$ and $n = 5$. Working similarly for the other two terms, we obtain

$$
\begin{aligned}
R_X = 2 & \frac{(2F_{ST} + (1 - F_{ST})p_A)(F_{ST} + (1 - F_{ST})p_B)}{(1 + 3F_{ST})(1 + 4F_{ST})} \\
& + \frac{(2F_{ST} + (1 - F_{ST})p_B)(F_{ST} + (1 - F_{ST})p_B)}{(1 + 3F_{ST})(1 + 4F_{ST})} \\
& + 2\frac{(F_{ST} + (1 - F_{ST})p_B)(F_{ST} + (1 - F_{ST})p_C)}{(1 + 3F_{ST})(1 + 4F_{ST})} \\
& = \frac{(F_{ST} + (1 - F_{ST})p_B)(8F_{ST} + (1 - F_{ST})(2p_A + p_B + 2p_C))}{(1 + 3F_{ST})(1 + 4F_{ST})}.
\end{aligned}
$$

All three possibilities for $\mathcal{G}_X$ involve alleles that have already been observed in the genotypes of either Q or K. Thus, possible coancestry has the effect of increasing[4] the probability of all possible $\mathcal{G}_X$ over the $F_{ST} = 0$ case.

In the above example, the CSP includes one allele not shared with K and, hence, which must have come from Q/X. There are essentially only two other cases: that zero and two alleles from Q/X can be determined by subtracting K's alleles from the CSP. An example of the former situation arises when both the CSP $= \mathcal{G}_Q = \mathcal{G}_K =$ AB. Then $\mathcal{G}_X$ can be any of AA, AB or BB. If instead, CSP = ABC and $\mathcal{G}_K =$ AA, then $\mathcal{G}_X =$ BC.

6.5.2.3   Two contributors: Q/X and U

Here, difficulties arise because of the number of scenarios to explain the components of the mixture. The relevant LR may be

$$
R_X = \frac{P(CSP | \mathcal{G}_Q;\ X \text{ and } U \text{ are the sources})}{P(CSP | \mathcal{G}_Q;\ Q \text{ and } U \text{ are the sources})}. \tag{6.14}
$$

The numerator requires summation over the conditional genotype probabilities, given $\mathcal{G}_Q$, of all $\mathcal{G}_X$ and $\mathcal{G}_U$ consistent with the CSP. The denominator requires summation over $\mathcal{G}_U$ only. Both summations need to take into account any relatedness among X, Q and U.

If two co-defendants, $Q1$ and $Q2$, are both alleged to have contributed DNA to the crime sample, then a court may be interested in the strength of evidence for, say, $Q1$

---

[4] The probability could decrease if one of p_A, p_B or p_C were much larger than 0.5.

to be a contributor of DNA to the crime sample both with and without assuming that $Q2$ is also a contributor, which could be addressed using the following two LRs:

$$R_{X2} = \frac{P(CSP|\mathcal{G}_{Q1}, \mathcal{G}_{Q2};\ X \text{ and } Q2 \text{ are the sources})}{P(CSP|\mathcal{G}_{Q1}, \mathcal{G}_{Q2};\ Q1 \text{ and } Q2 \text{ are the sources})}$$

$$R_{XU} = \frac{P(CSP|\mathcal{G}_{Q1}, \mathcal{G}_{Q2};\ X \text{ and } U \text{ are the sources})}{P(CSP|\mathcal{G}_{Q1}, \mathcal{G}_{Q2};\ Q1 \text{ and } U \text{ are the sources})}.$$

Computing each numerator and denominator of these LRs requires summing over all possible genotypes for whichever of X and U is included. Both observed genotypes $\mathcal{G}_{Q1}$ and $\mathcal{G}_{Q2}$ are potentially informative in both LRs due to possible coancestry with X.[5]

The two LRs can differ greatly in value. Intuitively, if $Q2$ is a source of the crime scene DNA, then he can explain many of the observed alleles, thus narrowing the possibilities for $\mathcal{G}_X$. In that case, R_{X2} would be larger than R_{XU}. However, if both deny being a source of the DNA, a court trying both men jointly may take the view that only R_{XU} can be used, since the presence of DNA from $Q2$ cannot be assumed when assessing the case against $Q1$, and vice versa. These are decisions for the court; the forensic scientist should try to foresee the reasonable possibilities for the pairs of hypotheses that the court may wish to compare.

Example

Suppose that, at a particular locus, the following are observed:

$$\mathcal{G}_{Q1} = AB \qquad \mathcal{G}_{Q2} = CC \qquad CSP = ABC.$$

The CSP is consistent with $Q1$ and $Q2$ being the sources of the DNA and with any other pair of individuals whose genotypes include alleles A, B and C only. Then

$$R_{XU} = \frac{P(ABC \mid \mathcal{G}_{Q1} = AB, \mathcal{G}_{Q2} = CC, X \text{ and } U \text{ are the sources})}{P(ABC \mid \mathcal{G}_{Q1} = AB, \mathcal{G}_{Q2} = CC, Q1 \text{ and } U \text{ are the sources})}. \qquad (6.15)$$

Its evaluation will depend on the relationships among $Q1$, $Q2$ and X. Here, for simplicity, we will assume that all are unrelated, and initially, we will also set $F_{ST} = 0$, so that $\mathcal{G}_{Q2}$ is irrelevant to R_{XU}.

The numerator of (6.15) is $12p_A p_B p_C (p_A + p_B + p_C)$. To see this, consider the event that it is the C allele that arises twice in the genotypes of X and U: this could result from a CC homozygote and an AB heterozygote (probability $4p_A p_B p_C^2$; one factor of two comes from the heterozygote, the other from the orderings of the two genotypes) or from an AC heterozygote and a BC heterozygote (probability $8p_A p_B p_C^2$), giving a total probability of $12p_A p_B p_C^2$. The expression for the numerator comes from combining this with the two terms corresponding to the A and the B alleles being represented twice.

---

[5] U may also have coancestry with $Q1$ or $Q2$, but the effect on numerator and denominator are similar, so the effect on the LR is typically negligible.

The denominator of (6.15) is the probability that $\mathcal{G}_U \in \{AC, BC, CC\}$, which is $p_C(2p_A + 2p_B + p_C)$, and so

$$R_X = \frac{12p_A p_B(p_A + p_B + p_C)}{2p_A + 2p_B + p_C}.$$

If $p_A = p_B = p_C = p$, then R_X reduces to $36p^2/5$, which takes minimum value 1.25 when $p = 1/3$: in this case, the evidence is of little value; however, if p is small the evidence is stronger: $R_X = 13.9$ if $p = 0.1$ and $R_X = 55.6$ if $p = 0.05$.

Now assume that $Q1$, $Q2$, X and U are all unrelated but are drawn from the same subpopulation for which the level of coancestry, relative to the population allele frequencies, can be characterised by a given value of F_{ST}. Now, the probability that it is the C allele that arises twice in the genotypes of X and U, given the observed genotypes of $Q1$ and $Q2$, is 12 times the probability that an ordered sample of size four is ABCC, given that a sample of size four has already been observed to be ABCC. Using the sampling formula (5.6) four times leads to

$$12\frac{(F_{ST} + (1-F_{ST})p_A)(F_{ST} + (1-F_{ST})p_B)(2F_{ST} + (1-F_{ST})p_C)(3F_{ST} + (1-F_{ST})p_C)}{(1 + 3F_{ST})(1 + 4F_{ST})(1 + 5F_{ST})(1 + 6F_{ST})}$$

and the remaining two terms of the numerator may be computed similarly. The denominator of R_X becomes

$$2\frac{(2F_{ST} + (1 - F_{ST})(p_A + p_B))(F_{ST} + (1 - F_{ST})p_C)}{(1 + 3F_{ST})(1 + 4F_{ST})}$$
$$+ \frac{(2F_{ST} + (1 - F_{ST})p_C)(3F_{ST} + (1 - F_{ST})p_C)}{(1 + 3F_{ST})(1 + 4F_{ST})}.$$

The final result for R_X is tedious to write down but easy to compute. For example, when $p_A = p_B = p_C = 1/3$, the value of 0.8 at $F_{ST} = 0$ increases only slightly to a maximum of 0.8007 at $F_{ST} \approx 1\%$ and then declines as F_{ST} increases. Thus the assumption of a large level of coancestry among $Q1$, $Q2$ and X in this setting *strengthens* the case against $Q1$. However, if $p_A = p_B = p_C = 0.1$, then increasing F_{ST} weakens the case against $Q1$: R_X increases from 0.072 at $F_{ST} = 0$ to 0.094 at $F_{ST} = 1\%$ and 0.17 at $F_{ST} = 5\%$.

6.5.3   Quantitative interpretation of mixtures

Gill et al. [2006] described a set of procedures, based on that proposed by Clayton et al. [1998] for interpreting two-contributor mixtures that involves using peak heights from the epg to estimate the heterozygote balance for each pair of alleles. Possible genotype configurations with extreme heterozygote imbalance are then excluded from consideration. For example, for the epg of Figure 6.2, we noted in Section 6.5.1 that the peak heights at several loci indicated four distinct alleles of which two, say alleles A and B, gave strong signals (with approximately equal peak heights) and two, say C and D, gave weak signals (also with approximately equal peak heights). The qualitative approach discussed above would count all three genotype pairs consistent with these alleles (AB,CD; AC,BD; and AD,BC) as equally plausible. The semi-quantitative approach of Clayton et al. [1998] would regard only the AB,CD genotype pair as plausible, because it pairs up the alleles with the two strong and the two weak signals. Congruent genotype combinations across loci are determined through estimating a mixture proportion for each contributor.

This semi-quantitative approach seems to be widely employed in practice, but it is not without difficulties. Estimating mixture proportion relies on an assumption that this proportion is approximately the same across loci. This seems a reasonable assumption prior to the polymerase chain reaction (PCR) step in the STR typing procedure, and indeed Gill et al. [1998] examined the estimation of mixture contribution proportions and found that consistency across loci is the norm. However, Bill et al. [2005] cited internal Forensic Science Service data, suggesting that single-locus mixture proportion estimates can vary by a factor of 0.35 compared to a global estimate, which can result in invalid inferences about plausible genotype pairs at these loci. Further, deciding which genotype configurations are consistent with the mixture proportions is difficult in some borderline cases. For further description and criticisms, see Buckleton et al. [2005].

As a result, Gill et al. [1998] went on to suggest a more quantitative approach in which a weight was given to each possible genotype allocation according to its plausibility, given the estimated mixture proportion. The ideal, fully quantitative approach to assess the weight of evidence for STR mixtures would involve all the information contained in the epg. Evett et al. [1998] set out a framework for analyses taking peak areas into account. The approach has been expanded on, with software in development or available from multiple authors utilising peak heights [Cowell et al., 2014b; Bright et al., 2013c; Puch-Solis et al., 2013b; Perlin et al., 2011b]. A consensus on the modelling assumptions for such analysis has yet to be reached, as this is an area of current active development.

6.6   Identification exercises

Solutions start on page 187.

6.1 Suppose that a man Q is accused of being the source of a crime sample. At two STR loci, his genotypes are AB and CC, respectively, in both cases matching the CSP. The crime occurred in an isolated village, and it is assumed that 1 of 45 men is guilty; these include Q, his half-brother h, two uncles, and 41 unrelated men. There has been no recent migration into the village, and the background level of coancestry among the men, relative to the wider population, is taken to be $F_{ST} = 3\%$.

   (a) Calculate the probability that Q is the culprit if all 45 men are initially regarded as equally under suspicion, and the allele proportions in the wider population are $p_A = 6\%$, $p_B = 14\%$ and $p_C = 4\%$.

   (b) Repeat (a), but now there is a camp of migrant labourers nearby and 20 men from this camp are each just as likely to be the source of the CSP as each of the 45 men in the village. The migrants come from a distant population in which $p_A = 8\%$, $p_B = 9\%$ and $p_C = 6\%$.

6.2 Suppose that you are the juror in a case that rests on the identification of a single contributor to a CSP via a DNA profile match. You have heard a forensic scientist report a match probability of 10^{-9} for an alternative culprit unrelated to Q. You accept this value and also that all relatives of Q are excluded from suspicion. However, you mistrust the island police and, based on the non-DNA evidence, you assess that there is a 1% chance that the police have maliciously swapped the crime sample with a sample taken directly from Q. How would you calculate the appropriate LR, taking into account both the possibility of a chance match and the possibility of fraudulent evidence?

6.3 Using the qualitative approach of Section 6.5.2, formulate single-locus LRs conveying the weight of evidence against a suspect Q given a mixed CSP in the following situations. Assume that there are exactly two contributors to the mixed profile and that all actual and possible contributors are unrelated but come from the same subpopulation characterised by a given value of F_{ST}. Express your answer in terms of F_{ST} and allele proportions such as p_A and p_B.

(a) CSP $=$ AB, and $\mathcal{G}_Q =$ AA. One of the contributors to CSP is known (K) and $\mathcal{G}_K =$ AB.

(b) CSP $=$ ABCD, and $\mathcal{G}_Q =$ AB. One of the contributors to the CSP is known, and $\mathcal{G}_K =$ CD.

(c) CSP $=$ ABCD, and $\mathcal{G}_Q =$ AB. The second contributor is unknown.

6.4 Consider again the example introduced on page 104 and assume that $F_{ST} = 0$.

(a) Work out an expression for an LR assessing the strength of the evidence against $Q2$.

(b) Suppose now that the epg at every locus is consistent with there being a strong imbalance in the contributions to the mixture from the two source individuals. At the locus in question, the peaks corresponding to alleles A and B are about the same height, whereas the peak for allele C is much lower.

   (i) What is the most likely genotype for the major contributor at this locus? Assuming this genotype, derive an appropriate LR assessing the hypothesis that $Q1$ is a contributor to the profile.

   (ii) What genotypes are possible for the minor contributor to the mixed profile? Derive an appropriate LR assessing the hypothesis that $Q2$ is a contributor to the profile.

7

Inferring relatedness

7.1 Paternity

7.1.1 Weight of evidence for paternity

The LRs for assessing an allegation that Q is the father of a child c, given that M is the mother, have the form

$$R_X = \frac{P(E \mid X \text{ and } M \text{ are the parents of } c)}{P(E \mid Q \text{ and } M \text{ are the parents of } c)}, \tag{7.1}$$

for each alternative possible father, X. In (7.1), E denotes all the evidence, but we will here ignore the non-DNA evidence (the principles are the same as in Section 3.4.2) and assume that E includes only the DNA evidence. Typically, this will consist of the genotypes $\mathcal{G}_M$, $\mathcal{G}_Q$ and $\mathcal{G}_c$, in which case we can write

$$\begin{aligned} R_X &= \frac{P(\mathcal{G}_c, \mathcal{G}_Q, \mathcal{G}_M \mid X \text{ is father})}{P(\mathcal{G}_c, \mathcal{G}_Q, \mathcal{G}_M \mid Q \text{ is father})} \\ &= \frac{P(\mathcal{G}_c \mid \mathcal{G}_Q, \mathcal{G}_M; \text{ father is } X)}{P(\mathcal{G}_c \mid \mathcal{G}_Q, \mathcal{G}_M; \text{ father is } Q)}. \end{aligned} \tag{7.2}$$

The second equality follows if, as seems reasonable, we assume that the probabilities of observing $\mathcal{G}_Q$ and $\mathcal{G}_M$ are unaffected by whether or not Q is the father of c. An LR for paternity is often referred to as a paternity index.

The general issues for interpretation of R_X in the paternity setting are the same as those for identification, discussed in Sections 3.2 and 3.3. The LRs for DNA evidence can be computed, as well as those for other evidence such as the statements of M and Q and information about their locations of residence and the frequency of contact between them. These LRs can then be combined using the weight-of-evidence

Weight-of-Evidence for Forensic DNA Profiles, Second Edition.
David J. Balding and Christopher D. Steele.
© 2015 John Wiley & Sons, Ltd. Published 2015 by John Wiley & Sons, Ltd.
Companion Website: www.wiley.com/go/balding/weight_of_evidence

formula (3.3) to arrive at an overall probability that Q is the father. In particular, just as for identification, this probability involves a summation over all alternative fathers X, including the close relatives of Q, not just a single 'random man' alternative father.

The weight-of-evidence formula is concerned with assessing the truth of the allegation and not directly with any subsequent decisions. Thus, the assessment of evidence for paternity based on the formula is in principle the same whether it is a criminal case or a civil dispute. These settings differ in the threshold required for decision making by the court: in many legal systems, a probability of paternity over 1/2 should suffice to prevail in a civil case (proof on the balance of probabilities), whereas for a criminal case, a probability close to 1 is needed for a satisfactory conviction (beyond reasonable doubt). Moreover, it is usually accepted that approximations should tend to favour Q in the criminal setting (see Section 6.3), whereas this is often not appropriate in a civil case.

7.1.2   Prior probabilities

The prior probability of an allegation against Q should reflect the actual circumstances whether it is a criminal or civil allegation. Many forensic scientists, lawyers and academic commentators seem reluctant to consider in the paternity setting the very low prior probabilities that are now often accepted for identification. Indeed, there is a shamefully high prevalence of an unjustified assumption that both Q and an unrelated 'random man' X have a prior probability 1/2 of being the father [Koehler, 1992]. This 'even prior' assumption is convenient, since it implies that the LR R_X is also the posterior odds against Q being the father. It may be based on the misguided assumption that equal prior weight should be assigned to the, typically conflicting, claims of Q and M. However, this ignores the reality of, for example, additional background evidence, and multiple alternative fathers, perhaps including relatives of Q.

Often, in paternity cases, Q is known to M; they may have a blood relationship and may live in proximity to each other in which case there are few if any alternative fathers in a similar situation to Q; these are matters for a court to evaluate. Some commentators have argued that considering, say, a 1000 alternative possible fathers is to slur M with the allegation that she had sexual intercourse with all of them. Others have tried to formulate prior probabilities on the basis of an estimate of the number of sexual partners of M. These lines of thought are both misguided: M may have had sex with 1 or a 1000 men, the number is irrelevant. We know that she had sexual intercourse with at least one man, the father of c, and there may well be millions of possibilities for the identity of that one man, even if M only had sex once in her life.

Attempting to assess a prior probability from the rate of exclusions in previous cases [Chakravarti and Li, 1984] is also misguided: background information relevant to the present case should be taken into account. Recommendations from the International Society for Forensic Genetics allow illustrative calculations with multiple prior probabilities but not a definitive prior, which is a matter for jurors [Gjertson et al., 2007]. If a probability of paternity is given, the priors used should be stated explicitly when reporting, and sensitivity analysis relating to the choice of prior is recommended.

7.1.3   Calculating LRs

We will initially assume that X, M and Q have no direct relatedness. Direct relationships between X and Q will be considered in Section 7.1.5 and between M and Q (incest) in Section 7.1.6. The situation in which $\mathcal{G}_M$ is not available is discussed in Section 7.1.7. We will also (until Section 7.1.8) ignore the possibility of mutation. This means that R_X is undefined if M does not share an allele with c at every locus and is zero if $\mathcal{G}_Q$ is inconsistent with Q and M being the parents of c.

Consider first a single locus, and assume that c's paternal allele can be determined (without assuming that Q is the father), either because c is homozygous or because she/he shares exactly one allele in common with M. In this case, a man is excluded as a possible father if his profile does not include c's paternal allele. Then the probabilities for transmission of the maternal allele cancel in (7.2) to obtain

$$R_X = \frac{P(\text{paternal allele of } c \mid \mathcal{G}_Q, \mathcal{G}_M; \text{father is } X)}{P(\text{paternal allele of } c \mid \mathcal{G}_Q, \mathcal{G}_M; \text{father is } Q)}. \qquad (7.3)$$

For example, we may observe the $\mathcal{G}_M$ and $\mathcal{G}_c$ shown in Figure 7.1, in which case we know that the child's paternal allele is C and any man lacking a C allele is excluded from being the father of c.

7.1.3.1   Ignoring F_{ST}

Ignoring coancestry ($F_{ST} = 0$), $\mathcal{G}_M$ plays no further role if it determines c's paternal allele. For the example of Figure 7.1, the denominator of (7.3) is 1 if $\mathcal{G}_Q = $ CC and 1/2 if $\mathcal{G}_Q = $ CF, for some F $\neq$ C. The numerator is the probability that an allele drawn from X is C. Since $\mathcal{G}_X$ is unavailable, this probability can be regarded as p_C, the proportion of C alleles in the population of X. Here we will assume that p_C is known without error; in practice, an estimate is obtained from a population database (see Section 6.3.1). Then

$$R_X = \begin{cases} p_C & \text{if } \mathcal{G}_Q = \text{CC} \\ 2p_C & \text{if } \mathcal{G}_Q = \text{CF}. \end{cases} \qquad (7.4)$$

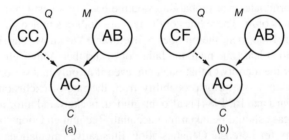

Figure 7.1   Single-locus genotypes in a case of questioned paternity. The solid arrows indicate the known mother–child relationship, and the dashed arrows indicate putative father–child relationships. Here, $\mathcal{G}_M = AB$ and $\mathcal{G}_c = AC$ so that, ignoring mutation, the father must have genotype (a) CC or (b) CF, for some F $\neq$ C.

If $p_C > 0.5$, then $R_X > 1$ in the heterozygote case, and so the DNA evidence points away from Q as a possible father, even though Q has c's paternal allele. This situation rarely arises in practice but illustrates how LRs can sometimes clarify valid points that may seem counter-intuitive. Li and Chakravarti [1988] criticized the use of LRs for establishing paternity, arguing that it was 'astonishing' that a hypothetical long sequence of such loci would suggest convincing evidence that Q is not the father, even though he is not excluded at any locus from being the father. This may seem counter-intuitive, but some reflection shows that it is correct: see Section 10.2.

7.1.3.2   Incorporating F_{ST}

The calculations leading to (7.4) ignore possible coancestry between X and one or both of Q and M. If X is assumed to have coancestry with Q, and possibly also M, measured by F_{ST}, then we can use the sampling formula (5.6) to compute the LR (7.3).

Continuing with the example of Figure 7.1, the denominator of (7.3) is once again either 1 or 1/2, according to whether Q is homozygous or heterozygous. In the numerator, we require the conditional probability of c's paternal allele, given the profiles of Q and M and the hypothesis that X is the father. Because of coancestry, the profiles of Q and M may be informative about the allele transmitted by X to c, and so these profiles cannot automatically be neglected as was done when $F_{ST} = 0$.

First, suppose that the coancestry of X and Q is specified by F_{ST}, but M has no coancestry with either (e.g. she is from a different population). Then G_M is uninformative about G_X, and (7.3) becomes, in the homozygous and heterozygous cases, respectively,

$$R_X = \begin{cases} P(C \text{ has paternal allele C} \mid G_Q = \text{CC, father is } X) \\ 2P(C \text{ has paternal allele C} \mid G_Q = \text{CF, father is } X). \end{cases} \qquad (7.5)$$

Since X is assumed to be the father, G_Q is not directly relevant in (7.5), but it is indirectly relevant because X's profile is unknown and he may share alleles with Q due to coancestry.

When $G_Q = \text{CC}$, since we are given that X is the father, the paternal allele of c may be regarded as a third random draw from the population, and so the required probability is that a third allele is again C, given that two observed alleles are both C. Invoking the sampling formula (5.6) with $m = n = 2$, we obtain

$$R_X = P(C \mid CC) = \frac{2F_{ST} + (1 - F_{ST})p_C}{1 + F_{ST}} \quad \text{if } G_Q = \text{CC},$$

Reasoning similarly in the heterozygote case, we have

$$R_X = 2P(C \mid CF) = 2\frac{F_{ST} + (1 - F_{ST})p_C}{1 + F_{ST}} \quad \text{if } G_Q = \text{CF}. \qquad (7.6)$$

These expressions are reported in the final column of Table 7.1 (the heterozygote case is shown in rows 4, 7, and 9, according to the possible values for F, but R_X is the same in each case).

Table 7.1   Single-locus LRs for paternity.

		Likelihood ratio $\times(1 + F_{ST})$		
$\mathcal{G}_Q$	$\mathcal{G}_M$	$\mathcal{G}_c = AA$	$\mathcal{G}_c = AB$	$\mathcal{G}_c = AC$
AA	AA	$2F_{ST}+(1-F_{ST})p_A$		
AB	AA	$2(F_{ST}+(1-F_{ST})p_A)$	$2(F_{ST}+(1-F_{ST})p_B)$	
BB	AA		$2F_{ST}+(1-F_{ST})p_B$	
BC	AA		$2(F_{ST}+(1-F_{ST})p_B)$	$2(F_{ST}+(1-F_{ST})p_C)$
AA	AB	$2F_{ST}+(1-F_{ST})p_A$	$2F_{ST}+(1-F_{ST})(p_A+p_B)$	
AB	AB	$2(F_{ST}+(1-F_{ST})p_A)$	$2F_{ST}+(1-F_{ST})(p_A+p_B)$	
AC	AB	$2(F_{ST}+(1-F_{ST})p_A)$	$2(F_{ST}+(1-F_{ST})(p_A+p_B))$	$2(F_{ST}+(1-F_{ST})p_C)$
CC	AB			$2F_{ST}+(1-F_{ST})p_C$
CD	AB			$2(F_{ST}+(1-F_{ST})p_C)$

The alleged father Q is assumed unrelated to X, but they have nonzero coancestry, measured by F_{ST}; both are assumed here to have no coancestry with the mother M. To simplify the presentation, the $1 + F_{ST}$ that divides every entry in the table has been omitted.

Suppose now that we assume that all three X, M and Q have a common level of coancestry F_{ST}. Then we have

$$R_X = \text{P(C} \mid \text{ABCC)} = \frac{2F_{ST}+(1-F_{ST})p_C}{1+3F_{ST}} \quad \text{if } \mathcal{G}_Q = \text{CC} \tag{7.7}$$

$$= 2\text{P(C} \mid \text{ABCF)} = 2\frac{F_{ST}+(1-F_{ST})p_C}{1+3F_{ST}} \quad \text{if } \mathcal{G}_Q = \text{CF.} \tag{7.8}$$

These, together with the formulae corresponding to other possible observed profiles, are given in Balding and Nichols [1995].

In both (7.6) and (7.8), we may have F = A or F = B; the result is unaffected (but recall that F $\neq$ C). Both heterozygote formulas reduce to $R_X = 2p_C$ if $F_{ST} = 0$, in agreement with (7.4), and similarly, the two formulas in the homozygous case reduce to $R_X = p_C$. Because Q carries a C allele, possible coancestry between X and Q usually makes it more likely that X also carries a C allele, which (slightly) lessens the evidential strength against Q. This effect becomes more important as C becomes rarer. The direction of the effect can be reversed – so that consideration of coancestry can strengthen the evidence against Q – in the heterozygote case if p_C is large (larger than actually arises in practice).

7.1.3.3   Paternal allele ambiguous

Consider the case of Figure 7.2(a) for which the LR is

$$R_X = \frac{\text{P}(\mathcal{G}_c = \text{AB} \mid \mathcal{G}_M = \text{AB}, \mathcal{G}_Q = \text{AA}, X \text{ is father})}{\text{P}(\mathcal{G}_c = \text{AB} \mid \mathcal{G}_M = \text{AB}, \mathcal{G}_Q = \text{AA}, Q \text{ is father})}. \tag{7.9}$$

The denominator takes value 1/2 irrespective of any assumptions about F_{ST}, because there is probability 1/2 that M transmits allele B to c.

Ignoring F_{ST}, if X is the father, then G_Q is irrelevant, and we must consider the two possible transmissions of an allele from M to c, each of which has probability $1/2$. In either case, c must have received the other allele from X, and hence, the LR is

$$R_X = \frac{(p_A + p_B)/2}{1/2} = p_A + p_B.$$

If now we assume coancestry between X and Q only, then R_X becomes

$$R_X = P(A \mid AA) + P(B \mid AA)$$
$$= \frac{2F_{ST} + (1 - F_{ST})p_A}{1 + F_{ST}} + \frac{(1 - F_{ST})p_B}{1 + F_{ST}}$$
$$= \frac{2F_{ST} + (1 - F_{ST})(p_A + p_B)}{1 + F_{ST}}.$$

This formula is given in row 5, middle column, of Table 7.1.

For the case of Figure 7.2b, the denominator of the LR is $1/2$, since there are two possible pairs of transmissions from M and Q to c, and each possibility has probability $1/4$. It then follows that R_X is the same as for Figure 7.2a:

$$R_X = P(A \mid AB) + P(B \mid AB)$$
$$= \frac{F_{ST} + (1 - F_{ST})p_A}{1 + F_{ST}} + \frac{F_{ST} + (1 - F_{ST})p_B}{1 + F_{ST}}$$
$$= \frac{2F_{ST} + (1 - F_{ST})(p_A + p_B)}{1 + F_{ST}},$$

given in row 6, middle column, of Table 7.1. The LR for the case of Figure 7.2c is given in row 7 of the table.

See Ayres and Balding [2005] for LRs given coancestry of all possible pairs from M, Q and X, Balding and Nichols [1995] for case that all three of c, M and Q have coancestry at the same level, and Fung et al. [2003] for some examples of the effect of an F_{ST} adjustment in paternity cases.

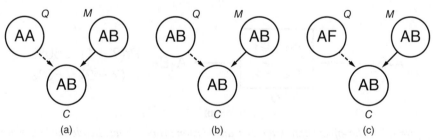

Figure 7.2   Single-locus genotypes in a case of questioned paternity in which G_M is available, but the child's paternal allele remains ambiguous. Three of the five possible paternal genotypes are shown, the remaining two are the same as (a) and (c) but replacing A with B. Here, F ≠ A and F ≠ B.

7.1.4   Multiple loci: the effect of linkage

For loci on different chromosomes, the single-locus LRs considered above can be multiplied to obtain whole-profile LRs; the issues are the same here as discussed in Section 6.2.3. However, two loci in the CODIS system are both on chromosome 5, separated by approximately 25 cM. This implies that alleles at these loci are co-inherited in about 80% of parent–child transmissions, much greater than the 50% co-inheritance that applies to unlinked loci. This co-inheritance probability is far enough from 100% that for unrelated individuals, it would not be expected to affect the validity of the 'product rule' (see Section 6.2.3), but for close relatives, the effect may be noticeable.

Linkage does not necessarily nullify the validity of the product rule. Suppose that Q is heterozygous at two linked loci, A and B, having genotypes A_1A_2 and B_1B_2. If he transmits A_1 to his child, what is the probability that B_1 will also be transmitted? This scenario is illustrated in Figure 7.3a. Because the loci are linked, whichever allele, B_1 or B_2, is in phase with A_1 is more likely to be transmitted with it. But this information is of no use if we are ignorant of the phase; in that case, the transmission probabilities are 50:50, just as in the unlinked case. This provides another reminder of the fact, discussed in Section 6.2.3, that the independence of two events is not an absolute state of nature but a function of what we currently know (or assume).

Ignorance of phase only protects from the effects of linkage in simple pedigrees, such as a family trio of parents and child, in which each allele is involved in only one transmission. If genotypes are available on multiple transmissions within a pedigree, the co-inheritance pattern of alleles in one transmission is potentially informative about phase and, hence, affects the probabilities at the same locus in another transmission. For example, knowing that the man described above received

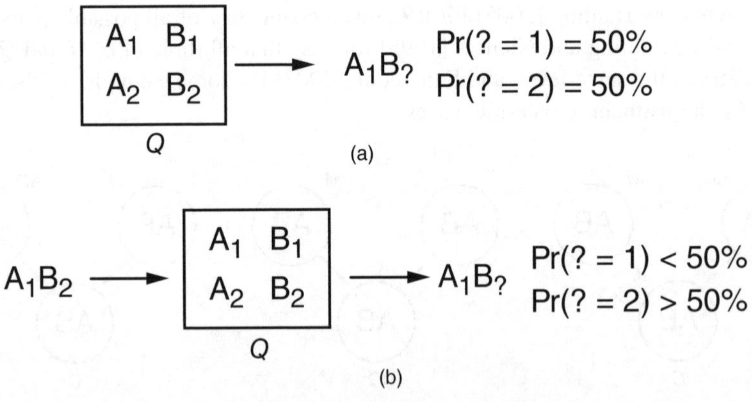

Figure 7.3   The effects of linkage on transmission probabilities. Inside the boxes are shown the genotypes of Q at loci A and B. Arrows denote the parent–child transmission of a single haplotype. In (a), because phase is unknown, B_1 and B_2 are equally likely to be transmitted to the child. In (b), Q received the A_1B_2 haplotype from his father, providing information on phase, and so B_2 is more likely than B_1 to be transmitted with A_1.

A_1 and B_2 from his father makes it more likely that these two alleles are in phase and, hence, more likely that either both or neither will be transmitted to the man's child (Figure 7.3b). For a numerical example, see Buckleton et al. [2005].

Neglecting linkage may tend to overstate the evidence for paternity, though any effect is small if the loci are widely spaced (say, >25 cM). Removing one locus from a linked pair usually leads to a larger error but in the direction of understating the evidence. Linkage can be accounted for by calculating a joint likelihood for the linked loci, rather than multiplying likelihoods for the individual loci. The joint likelihood depends on the genotypes of the two individuals at both loci, and the genetic distance between the loci. The details are complex but essentially involve summing likelihoods conditional on each of nine possible identical by descent (IBD) states: $L_{x,y}$, where $x \in \{0, 1, 2\}$ and $y \in \{0, 1, 2\}$ indicate the numbers of alleles shared IBD between the two individuals at the first and second linked loci. See Bright et al. [2013b] for further details. Each term in the likelihood includes the dependence between IBD states due to linkage, in terms of the recombination rate between the linked loci. Recombination rates between commonly used linked loci can be found in Phillips et al. [2012] and Wu et al. [2014].

7.1.5 Q may be related to c but not the father

So far, we have assumed no direct relationship between Q and X, but in practice, we must often assess the possibility that Q is in fact a relative, such as a brother, of a true father X, and hence, Q is related to c but not the father. We write $\bar{\kappa}$ for the probability that an allele chosen at random from X is IBD with one or other allele of Q. Some values of $\bar{\kappa}$ for particular relationships were given in Table 6.3.

Consider again the case that the paternal allele of c can be determined, say C, and that $G_Q = $ CF. The denominator of the LR (7.3) is once again 1/2. A child of X can receive a paternal C allele in two ways:

- a copy of allele C that is IBD with the C allele of Q (probability $\bar{\kappa}/2$);

- an allele that is not IBD with either allele of Q (probability $1 - \bar{\kappa}$) but happens to be allele C (probability p_C).

Thus the LR corresponding to (7.3) in the case that X and Q are related is

$$R_X = \frac{\bar{\kappa}/2 + (1 - \bar{\kappa})p_C}{1/2} = \bar{\kappa} + (1 - \bar{\kappa})2p_C, \tag{7.10}$$

which reduces to the heterozygote case of (7.4) when Q and X are unrelated ($\bar{\kappa} = 0$). At the other extreme, if X and Q are identical twins, $\bar{\kappa} = 1$ and (7.10) give $R_X = 1$, corresponding to the fact that DNA evidence is of no value in distinguishing identical twins. For siblings, $\bar{\kappa} = 1/2$, and we obtain $R_X = 0.5 + p_C$. Similar to the unrelated case, discussed in Section 7.1.3, if $p_C > 0.5$, then the observation that Q is heterozygous favours his brother X as the father, rather than Q himself.

Equation (7.10) has the form

$$R_X = \bar{\kappa} + (1 - \bar{\kappa})R_X^u, \tag{7.11}$$

where R_X^u is the single-locus LR that applies when X and Q are unrelated. In fact, (7.11) is completely general, applying whatever the genotypes of Q, M and c (see Balding and Nichols, 1995). To understand why, consider the paternal allele of c under the hypothesis that X is the father:

- with probability $\bar{\kappa}$, it is IBD with an allele of Q, in which case the situation is the same as when Q is the father, and so the LR is 1;

- with probability $1 - \bar{\kappa}$, it is not IBD with an allele of Q, in which case the situation is the same as when X and Q are unrelated, for which the LR is R_X^u.

In particular, (7.11) applies whether or not F_{ST} is taken into account and so can be used in conjunction with Table 7.1, or the analogous expressions from Balding and Nichols [1995], to derive LRs that take into account both direct and indirect relatedness of X with Q.

7.1.6 Incest

It may arise that Q is acknowledged to be the father of M but is also accused of being the father of her child c (see Figure 7.4). Provided that the profiles of c, M and Q are consistent with the assertions that M is the mother of c and Q is the father of M, the effects of both these relationships cancel in the LR and they do not affect the calculation. Thus the LR comparing the putative relationship shown in Figure 7.4 with the alternative that an unrelated man X is the father of c is the same as that calculated in a non-incest setting in Section 7.1.3.

If Q is the father of both M and c, then with probability 1/2, he transmits the same allele to both of them and, also with probability 1/2, M transmits her maternal allele to c. Thus, with probability 1/4, c has both alleles IBD with her mother. An example of this is shown in Figure 7.4, where c has the same genotype as M which, if Q is the father of both, has arisen because they both received the A allele from Q. When M and c have the same heterozygous genotype, the paternal allele of c cannot be assigned under the proposition that an unrelated random man is the father.

Although the LR may be unaffected by incest, an appropriate prior probability may be especially difficult to specify, see the discussion in Section 7.1.2. It may also be important to consider other male relatives of M in addition to Q.

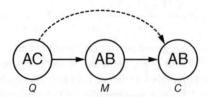

Figure 7.4   Relationships in an incest case. Circles represent individuals, solid arrows indicate accepted parent–child relationships and the dashed arrow indicates the putative relationship in question. Single-locus genotypes consistent with the putative relationship are also shown.

7.1.7 Mother unavailable

If $\mathcal{G}_M$ is unavailable, the appropriate version of the LR is

$$R_X = \frac{P(\mathcal{G}_c \mid \mathcal{G}_Q; \text{father is } X)}{P(\mathcal{G}_c \mid \mathcal{G}_Q; \text{father is } Q)}. \tag{7.12}$$

For example, we may require

$$R_X = \frac{P(\mathcal{G}_c = AA \mid \mathcal{G}_Q = AA; \text{father is } X)}{P(\mathcal{G}_c = AA \mid \mathcal{G}_Q = AA; \text{father is } Q)}.$$

Allowing for coancestry between Q and X, the denominator is

$$P(\mathcal{G}_c = AA \mid \mathcal{G}_Q = AA; \text{father is } Q) = \frac{P(AAA)}{P(AA)} = \frac{2F_{ST} + (1 - F_{ST})p_A}{1 + F_{ST}},$$

whilst the numerator is

$$\frac{P(AAAA)}{P(AA)} = \frac{P(AAA)}{P(AA)} \times \frac{3F_{ST} + (1 - F_{ST})p_A}{1 + 2F_{ST}},$$

and so

$$R_X = \frac{3F_{ST} + (1 - F_{ST})p_A}{1 + 2F_{ST}}.$$

LRs for this and other cases have been given by Ayres [2000b] and are shown in Table 7.2. Ayres and Balding [2005] gave LRs for the coancestry of all possible pairs of M, Q and X in the M-unavailable case (note that coancestry of Q or X with M can have an effect, even if M is not typed, since one of her alleles at each locus is observed in c).

7.1.8 Mutation

Mutation rates for STR loci are around 2 per 1000 generations, and so if an STR profile consists of 15 loci, there is roughly a 3% probability that there will be a mutation

Table 7.2 Single-locus LRs for paternity when $\mathcal{G}_M$ is unavailable.

c	Q	$R_X \times (1 + 2F_{ST})$
AA	AA	$3F_{ST} + (1 - F_{ST})p_A$
AA	AB	$2(2F_{ST} + (1 - F_{ST})p_A)$
AB	AA	$2(2F_{ST} + (1 - F_{ST})p_A)$
AB	AC	$4(F_{ST} + (1 - F_{ST})p_A)$
AB	AB	$4(F_{ST} + (1 - F_{ST})p_A)(F_{ST} + (1 - F_{ST})p_B)/(2F_{ST} + (1 - F_{ST})(p_A + p_B))$

The alleged father Q is assumed unrelated to X, but they have coancestry measured by F_{ST}. The denominator of each entry in the table is $1 + 2F_{ST}$, which has been removed to simplify the presentation. These formulas were first reported in Table 1 of Ayres [2000b].

Table 7.3   Three sets of genotypes
for which a mutation is required to
sustain the hypothesis that M and Q
are the parents of c.

Scenario		(i)	(ii)	(iii)
G_c	=	AB	AB	AB
G_M	=	AC	AC	DD
G_Q	=	CD	AD	AC

in transmission from father to child, in which case LRs that ignore possible mutations would imply a false exclusion.

The question arises as to whether, given profiles that are consistent with Q being the father of c at many loci, but an apparent exclusion at just one or perhaps two loci, it might be possible to report strong evidence that Q is the father and that one or more mutations have occurred. To answer this question quantitatively, we need to compute the appropriate LRs of the form (7.2) but now allowing for mutation. Dawid et al. [2001] gave a systematic treatment of this question when G_M is available, but ignoring coancestry. The latter omission is rectified by Ayres [2002], who treats the case that any two of M, Q and X have the same level of coancestry, F_{ST}, correcting formulas given in Ayres [2000b] that neglected the possibility of maternal transmission. Here, we work through three particular examples, shown in Table 7.3, and briefly discuss some additional issues not considered by these authors.

First, consider a general mutation model in which any allele can be transformed into any other allele in transmission from parent to child. We write $G_c = (C_1, C_2)$, $G_M = (M_1, M_2)$ and $G_Q = (Q_1, Q_2)$. Since any mutation is possible, under the hypothesis that Q is the father, C_1 could have descended from any one of the four parental alleles, and for each of these, there are two possibilities for the parental allele from which C_2 has descended. Four of these eight possibilities for the pair of transmissions are shown in Figure 7.5a. Thus, there are in principle eight terms contributing to the numerator of (7.2). However, since mutations are rare, any term corresponding to one (respectively, two) mutation(s) can in practice be neglected if there is also at least one term corresponding to zero (respectively, one) mutation.

Since the alternative father X is unprofiled, any mutation in transmission from X to c is unobserved and can be ignored, and the numerator of (7.2) has the four terms corresponding to the possible maternal transmissions indicated in Figure 7.5b. If c has at least one allele in common with M, then terms involving a mutation can be neglected.

7.1.8.1   Table 7.3, case (i)

Except for relabelling of alleles, this case is the same as that shown in row 15 of Table 1 of Dawid et al. [2001]. The genotypes are consistent with c having received A from M, but c has no allele in common with Q. The transmissions consistent with the minimal number of mutations under the hypotheses that (a) Q is the father and

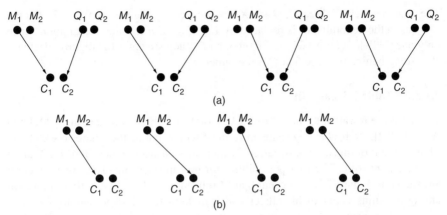

Figure 7.5   (a) Four of the eight possible transmissions of alleles from M and Q to c. The other four are obtained by interchanging C_1 with C_2. (b) The four possibilities for the transmission of an allele from M to c.

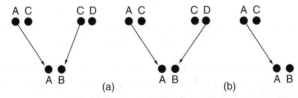

Figure 7.6   Allele transmissions for case (i) of Table 7.3. (a) Under the hypothesis that Q is the father, the two possible transmissions of alleles from M and Q to c consistent with one mutation. (b) The one possible transmission of an allele from M to c consistent with no mutation.

(b) an unprofiled man X is the father are shown in Figure 7.6. A factor of 1/2 for the transmission from M to c of allele A cancels in both numerator and denominator of (7.2). Ignoring this term, and F_{ST}, the numerator is p_B, whilst the denominator is $(\mu^f_{C\to B} + \mu^f_{D\to B})/2$, in which we introduce $\mu^f_{I\to J}$ to denote the probability that allele I mutates to allele J in transmission from father to child. Thus, the LR is

$$R_X = \frac{\text{P(paternal allele of } c \text{ is B} \mid \mathcal{G}_Q = \text{CD, father is } X)}{\text{P(paternal allele of } c \text{ is B} \mid \mathcal{G}_Q = \text{CD, father is } Q)}$$

$$= \frac{2p_B}{\mu^f_{C\to B} + \mu^f_{D\to B}}.$$

Assuming now that X and Q have coancestry at level F_{ST}, but each has no coancestry with M, the probability that an allele drawn from X is B, given that $\mathcal{G}_Q = \text{CD}$, is $(1 - F_{ST})p_B/(1 + F_{ST})$ and so

$$R_X = \frac{2(1 - F_{ST})p_B}{(1 + F_{ST})(\mu^f_{C\to B} + \mu^f_{D\to B})}.$$

The case in which all three M, Q and X have coancestry is given in Ayres [2002]: the $1 + F_{ST}$ in the denominator is replaced by $1 + 3F_{ST}$. Under either assumption, since only one B allele is observed, the effect of an F_{ST} adjustment is to (slightly) decrease R_X, strengthening the case that Q is the father.

7.1.8.2   Table 7.3, case (ii)

This case is equivalent to that of row 13 of Table 1 of both Dawid et al. [2001] and Ayres [2002]. Both M and Q share an A allele with c, but neither shares c's B allele. Thus a mutation must have occurred in transmission from either M or Q, if these are the parents of c. The four possibilities for transmissions involving one mutation are shown in Figure 7.7. The event that M transmits allele A to c with no mutation and Q transmits either of his alleles and it mutates to B has probability $(\mu^f_{A \to B} + \mu^f_{D \to B})/4$, while the event that Q transmits allele A to c with no mutation and M transmits either of her alleles and it mutates to B has probability $(\mu^m_{A \to B} + \mu^m_{C \to B})/4$, where the superscript m denotes that the mutation rate is for maternal transmission (typically lower than the rate for paternal transmissions). The numerator of (7.2) is $p_B/2$, and so

$$R_X = \frac{2p_B}{\mu^m_{A \to B} + \mu^m_{C \to B} + \mu^f_{A \to B} + \mu^f_{D \to B}}.$$

The effect of an F_{ST} adjustment is to decrease R_X by the same factor as in case (i), that is, either $(1 - F_{ST})/(1 + F_{ST})$ if just the coancestry of Q and X is considered or $(1 - F_{ST})/(1 + 3F_{ST})$ if M also has coancestry with Q and X.

7.1.8.3   Table 7.3, case (iii)

Here, Q shares an A allele with c, but M has no allele in common with c and so is excluded as the mother unless a mutation has occurred. If such a pattern arises at two or more loci, we should seriously question the proposed maternal relationship. However, if this relationship is accepted, then at least one mutation must have occurred, and as usual, we neglect the possibility of more than one mutation. Thus, one of the D alleles of M must have mutated to B in transmission to c; the two possibilities are indicated in Figure 7.8. These possibilities are the same irrespective of whether Q or

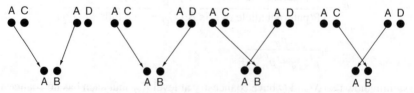

Figure 7.7   Allele transmissions for case (ii) of Table 7.3 under the hypothesis that Q is the father. The four possible transmissions of alleles from M and Q to c consistent with one mutation.

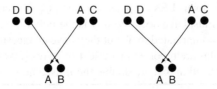

Figure 7.8   Allele transmissions for case (iii) of Table 7.3 under the hypothesis that Q is the father. The two possible transmissions of alleles from M and Q to c consistent with one mutation.

X is the father, and so the terms corresponding to them cancel in the LR, which is the same as in the no-mutation case when G_M is unavailable, given in row 4 of Table 7.2. Thus, the non-match of an allele between M and c has essentially no effect on the LR, except that it implies that the paternal allele of c is ambiguous.

7.1.8.4   Mutation models

We noted in Section 5.1.2 that not all STR mutations are equally likely. Single-step mutations, in which allele length changes by one repeat unit, are more frequent than multi-step mutations. A simple model that approximates this reality is the stepwise mutation model (SMM), in which both single-step mutations are assigned probability μ, whereas multi-step mutations have probability zero. Thus, under the SMM, Q is regarded as excluded from being the father if at any locus a multi-step mutation is required to sustain that hypothesis.

Dawid et al. [2001] discussed the SMM and more general mutation models. These authors gave an example of 13-locus profiles of M, Q and c, indicating exclusions of Q as the father at two loci. They calculate an overall LR (not allowing for coancestry) that indicates weak evidence for the non-paternity of Q. However, if one of the loci showing exclusions had been omitted, the LRs would indicate moderately strong evidence for paternity, despite the one remaining exclusion.

7.1.8.5   Mother unavailable

LRs in the case of apparent exclusion of the alleged father Q when G_M is unavailable are shown in Table 7.4. The values differ by a factor of two from those reported by Ayres [2000b], as Ayres used bidirectional mutation rates, whereas we give unidirectional mutations.

7.2   Other relatedness between two individuals

7.2.1   Only the two individuals profiled

If the two individuals are genotyped at a single locus, and assuming that no profiles of their relatives are available, the LR comparing a specified regular relationship

Table 7.4   Single-locus LRs for paternity when $\mathcal{G}_M$ is unavailable and when a mutation must have occurred if Q is the father of c. The alleged father Q is assumed unrelated to X, but they have coancestry, measured by F_{ST}. These formulas are reported in Table 4 of Ayres [2000b], but the mutation rates M used in that are double the μ used here.

c	Q	$R_X \times (1 + 2F_{ST})$
AA	BB	$(F_{ST} + (1 - F_{ST})p_A)/\mu^f_{B \to A}$
AA	BC	$2(F_{ST} + (1 - F_{ST})p_A)/\left(\mu^f_{B \to A} + \mu^f_{C \to A}\right)$
AB	CC	$2(1 - F_{ST})p_A p_B/\left(\mu^f_{C \to A} + \mu^f_{C \to B}\right)$
AB	CD	$4(1 - F_{ST})p_A p_B/\left(p_A\left(\mu^f_{C \to B} + \mu^f_{D \to B}\right) + p_B\left(\mu^f_{C \to A} + \mu^f_{D \to A}\right)\right)$

with the hypothesis that the individuals are unrelated is (here small values support no relatedness):

$$R = \kappa_0 + \kappa_1/R^p_X + \kappa_2/R^u_X, \tag{7.13}$$

in which the κ_j are the relatedness coefficients specifying the proposed relationship (defined at Table 6.3), and

- R^p_X denotes the single-locus LR for paternity in the unrelated case when $\mathcal{G}_M$ is unavailable; values are given in Table 7.2 when no mutations are required to sustain the parent–child hypothesis and in Table 7.4 when mutations are required (in practice, we can set $R^p_X = 0$ in these cases if $\kappa_1 < 1$).

- R^u_X denotes the single-locus LR for identity in the unrelated case. When the genotypes of the two individuals match, this is the match probability given at (6.4) and (6.5). Otherwise we interpret R^u_X as zero, in effect ignoring mutation, which would be important only in the case of identical twins, $\kappa_2 = 1$.

Explicit formulas corresponding to (7.13), and neglecting mutations, when

- $\kappa_1 = 1/2$, $\kappa_2 = 1/4$ (sibling),

- $\kappa_1 = 1/2$, $\kappa_2 = 0$ (half-sib, uncle/nephew, grandparent/grandchild),

- $\kappa_1 = 1/4$, $\kappa_2 = 0$ (cousin),

are given in Table 3 of Ayres [2000b]. Fung et al. [2003] also gave general expressions and explicit examples. Brenner and Weir [2003] described particular problems arising in the application of STR profiles to identify victims of the New York World Trade Center disaster in 2001.

7.2.2   Profiles of known relatives also available †

When a putative relationship between two individuals is being investigated via their DNA profiles, it can be helpful, when available, to have the profiles of known relatives of one or both of the individuals. For example, if the putative relationship is half-sibling through a common father, then knowing the profiles of the two mothers (here assumed unrelated) can strengthen the evidence for or against a common father. Perhaps surprisingly, when coancestry can be neglected (i.e. $F_{ST} = 0$), there are only four distinct forms for the LR:

(i) It can be verified that the children do not share a paternal allele, for example, child 1: $\mathcal{G}_{c1} = AB$, child 2 $\mathcal{G}_{c2} = CD$. In these cases the LR is 1/2.

(ii) Both paternal alleles can be identified and are the same, or the paternal allele of one child can be identified and is the same as one of the other child's alleles. An example of such genotypes, and the corresponding LR, is shown in row (ii) of Table 7.5.

(iii) Neither paternal allele can be identified, and the children have the same (heterozygous) genotypes. See row (iii) of Table 7.5.

(iv) Neither paternal allele can be identified, and the children are both heterozygous but share only one allele. See row (iv) of Table 7.5.

For highly polymorphic loci, case (i) applies frequently when the children are unrelated, whereas without the mothers' genotypes an LR of 1/2 is much less frequent. Although inclusion of the mothers' profiles can strengthen the evidence, the 10–15 STR loci currently in routine forensic will rarely suffice to convincingly establish a relationship as distant as half-sibling.

7.2.3 Software for relatedness analyses

Familias computes likelihoods and, hence, posterior probabilities for arbitrary pedigrees, given DNA data from some of the individuals. It permits various DNA

Table 7.5 Examples of single-locus LRs, ignoring F_{ST}, comparing the hypothesis that two children have a common father with that of unrelated fathers, with and without the genotypes of the children's mothers (assumed unrelated).

	Genotypes		Likelihood ratio	
	Mother	Child	With mothers	Ignoring mothers
(i)	AB AB	AC AD	$\frac{1}{2}$	$\frac{1}{2}\left(1 + \frac{1}{4p_A}\right)$
(ii)	AC AC	AB AB	$\frac{1}{2}\left(1 + \frac{1}{p_B}\right)$	$\frac{1}{2}\left(1 + \frac{1}{4p_A} + \frac{1}{4p_B}\right)$
(iii)	AB AB	AB AB	$\frac{1}{2}\left(1 + \frac{1}{p_A+p_B}\right)$	$\frac{1}{2}\left(1 + \frac{1}{4p_A} + \frac{1}{4p_B}\right)$
(iv)	AB AC	AB AC	$\frac{1}{2}\left(1 + \frac{p_A}{(p_A+p_B)(p_A+p_C)}\right)$	$\frac{1}{2}\left(1 + \frac{1}{4p_A}\right)$

Large values of the LR support the half-sibling relationship.

data types and mutation models, incorporates an F_{ST} adjustment for coancestry and is freely available for Microsoft platforms, together with documentation, from http://familias.name. Its use is also described in Egeland et al. [2000], who discussed selecting the most probable pedigrees to consider, when there are too many possible pedigrees to compute likelihoods for all of them. Another set of programs for paternity analysis is described in Fung [2003].

Bayesian networks (BN), also known as probabilistic expert systems, form a flexible class of computer software which implements Bayesian statistical models in such a way that the probability distributions of unknowns can be efficiently computed (via Bayes' theorem). Moreover, the probability distributions are readily updated when additional information is incorporated. The statistical model underlying a BN can be represented graphically: data and variables are represented by nodes, and these may be connected by directed arcs (arrows) signifying dependence relationships. Intuitively, an arc from node B to node C indicates that information about the value of B is potentially informative about the value of C. Informally, C is said to be the 'child' of B, and conversely, B is the 'parent' of C. Figure 7.9 illustrates two directed graphs in which B is the parent of both C and D.

Given the values of all the parents of a given node in the graphical representation of a BN, the value at that node is independent of the values at all nodes other than its descendants. For the graphs of Figure 7.9, D is conditionally independent of C given B; in Figure 7.9a, D is also conditionally independent of A given B, but this is not so for Figure 7.9b. This independence is *conditional*, any two of the variables are dependent unconditionally. Highlighting this conditional independence structure is the key to the computational efficiency of BN for many complex problems, since it allows them to be broken down into components. The definition of BN requires the underlying directed graph to be acyclic, which means that it has no paths (followed in the direction of the arcs) that end where they started. Equivalently, no node can be its own descendant. The 'loop' of Figure 7.9b is allowed because of the direction of the arcs (e.g. there is no arc away from A).

BNs are potentially useful for analysing problems of unknown relationships. Their flexibility permits, in principle, multiple individuals with some relationships known, additional background information, missing data and mutation. The language of 'parent' and 'child' suggests that the pedigree could serve as the directed, acyclic graphs on which an appropriate BN could be based. This would be the case for haploid species. For diploids, although a valid BN can be specified at the level of a pedigree, the need to separately track the inheritance of an individual's two alleles at

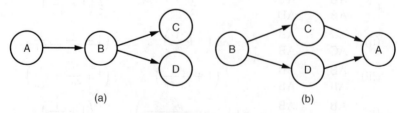

(a)   (b)

Figure 7.9   Two directed, acyclic graphs, each with four nodes.

an autosomal locus means that it is typically more useful to specify a richer graphical structure in which nodes correspond to alleles rather than individuals (see Lauritzen and Sheehan [2003], for a further discussion and an excellent, brief introduction to BN applied in genetics). As for all statistical modelling, there is no one 'correct' BN for the analysis of a relatedness problem: different representations of the problem can have different advantages or reflect different assumptions.

Dawid et al. [2002] gave several examples of BNs specified for the assessment of different forensic problems involving relatedness. They do not consider the incorporation of an F_{ST}-correction to allow for coancestry. This would add considerably to the complexity of the BN, because the alleles for which no parent is included in the model would no longer be independent.

Beyond relatedness problems, BNs have proved useful for many forensic applications involving a wide range of evidence types. While the difficulty of adapting general-purpose BN software to forensic applications initially limited their use, the development of object-oriented BNs [Bangsø and Wuillemin, 2000] allowed for development of forensic 'modules' that could then be combined by the user [Dawid et al., 2007], allowing for much simpler implementation. These object-oriented BNs have allowed for applications in paternity testing allowing for allelic dependencies [Hepler and Weir, 2008], weight-of-evidence evaluation for DNA evidence [Cowell et al., 2014b] and familial database search strategies [Cavallini and Corradi, 2006].

For a comprehensive introduction to BNs for general forensic applications see Taroni et al. [2014]. A review by Biedermann and Taroni [2012] covers the basics of BNs through their applications for DNA evidence.

7.3   Familial search

A national DNA database, such as the UK NDNAD, can be used to search for partial profile matches that may indicate relatedness between the DNA source and a member of the database [Bieber et al., 2006]. The first prosecution arising from identification of a relative through a partial DNA profile match occurred in Surrey, UK, in April 2004. The convicted man had no criminal record and no match of the CSP was found in the UK national DNA database. However, a partial match, of 16 alleles out of 20, was found, and this led police to investigate the immediate family of the individual providing the partial match. A full DNA profile match was obtained with a brother, who was charged and eventually convicted.

Currently, in the United Kingdom, possible parent–offspring relationships are identified on the basis of matching at least one allele per locus, and potential siblings are obtained by ranking database members by the number of matching alleles [Maguire et al., 2014]. Although it should be clear to forensic DNA experts that an LR such as (7.13) provides the best basis for the ranking, and this has been confirmed in simulation studies [Bleka et al., 2014; Balding et al., 2013], valuable information has been lost by preferring the simpler metrics based on matching allele counts, without taking account of the allele frequencies, number of contributors and degradation. A criterion based on matching allele count has even been integrated into state law in California [Rohlfs et al., 2013].

Since the general population contains many people unrelated to a given individual, there is potentially a problem that, by chance, some unrelated individuals resemble the crime scene DNA profile as closely as would be expected from a true relative. This is known as the 'multiple testing' problem in classical statistics, but this is a misnomer since the problem is not the multiplicity of the tests (i.e. the number of individuals searched in the database), rather it is that, in a large population, most pairs of individuals are unrelated. Thus, close relatives are a priori unlikely, and so substantial evidence is needed to be convinced of a sibling relationship. Appropriate prior probabilities can be based on average population values. For example, in a population of 5 million adult males in which the average number of brothers of any man is 0.5, the prior probability that two given men are brothers could be taken to be 1 in 10 million. Therefore, standard autosomal STR profiles of 15–20 loci are not sufficient to distinguish close relatives with high confidence, so familial searches usually return a long list of possible relatives.

Concerns that this practice brings many innocent individuals under police scrutiny have been raised, and if some sections of the population are already over-represented in the database, they will also be over-represented in investigations arising from familial searches [Rohlfs et al., 2013]. These authors also showed that a more distant relationship may often be mistaken for a first-degree relationship, so that the right family may come under scrutiny but the wrong members of it. Rather than only contrast the sibling relationship with unrelated, a more satisfactory approach is to use (7.13) to calculate a probability distribution for all the possible relationships of two profiled individuals (see Egeland et al., [2000]).

One way to refine the list of suspects is to type further markers, including the Y chromosome or mtDNA. This requires that biological samples from database members are retained for further analyses, which is generally no longer possible in the United Kingdom under new regulations [UK Parliament, 2013]. As of 2014 familial database searching has been used in 210 cases in the United Kingdom, 46 of which led to identification of a relative of the perpetrator [Maguire et al., 2014]. Familial database searching is also used in the Netherlands, New Zealand and some states of the United States and Australia [Maguire et al., 2014].

7.4   Inference of ethnicity †

When a DNA profile is obtained from a crime scene, it may be of assistance to investigators to predict the ethnic origin of its unknown source. Such predictions are complicated by the fact that 'ethnicity' and 'race' largely reflect social context rather than genetics, and 'ethnic' groups are not precisely defined. In particular, there are different possible levels of classification; for example, 'African', 'Nigerian' and 'Ibo' may all be applied to the same individual. Although different human groups have obvious phenotypic differences, these correspond to only a small part of the genome, and for typical loci, humans form a remarkably homogeneous species at the DNA level, and human allele frequencies usually vary smoothly with distance, with few sharp changes caused by geographic or social barriers.

For these reasons, and perhaps also because of sensitivity to social and political concerns, many authors have argued that notions of 'ethnicity' and 'race' have no useful role in genetics-related science. An alternative view is that in most nations, there are recognisable, distinct groups into which most of the population can reliably be allocated, for example, by self-report. These groups often have enough genetic differences that classification of a query sample of DNA into its source group can be performed with reasonable accuracy: sufficient to be of some use to crime investigators, provided that the limitations of the approach are appreciated.

Suppose that the population can be classified into K groups. Then, by the application of Bayes' theorem (Section 3.5), the posterior probability that a DNA sample with profile D comes from group k can be written as

$$P(\text{group } k \mid D) = \frac{P(D \mid \text{group } k)\pi_k}{\sum_{j=1}^{K} P(D \mid \text{group } j)\pi_j}, \tag{7.14}$$

in which π_j denotes the prior probability of group j. Population proportions of the different groups may provide acceptable values for the π_j, or it may be possible to use more specific local information, or possibly other information on the different patterns of offending recorded for the different groups.

If we assume Hardy–Weinberg equilibrium (HWE) (Section 5.4) and linkage equilibrium (Section 5.5), then each likelihood $P(D \mid \text{group } j)$ can be expressed in terms of a product over allele proportions in group j [Rannala and Mountain, 1997]. Of course, true allele proportions in the different groups will not be known, but these can be estimated from samples. A problem arises for rare alleles with a low sample count, since small differences attributable to stochastic variation could then lead to big differences in the likelihood used. Rannala and Mountain [1997] in effect employed a 'pseudo-count' approach to overcoming this problem, in which each sample count was incremented by the inverse of the number of distinct alleles at the locus. We have advocated a similar approach in Section 6.3.1, except that we have employed a pseudo-count of one.

Provided that the rare-allele problem is overcome, the above approach can give some information about the ethnic origin of an unknown offender from the STR profile. For example, a forensic scientist may be able to advise the police that *provided that* the offender (who left a DNA stain at the crime scene) comes from one of the major populations (such as 'European', 'Asian' and 'African'), she/he evaluates probabilities of 88% that the offender is of European origin, 9% of Asian origin and 3% of African origin. Because of the limited discriminatory power of forensic STR loci, it will be unlikely that a small minority group will be correctly identified as the source population.

Rather than relying on forensic STRs, single nucleotide polymorphism (SNP) profiles can be used for ethnic classification [Yang et al., 2005; Phillips et al., 2007; Halder et al., 2008; Jia et al., 2014]. This often uses SNP sets that have been ascertained to be useful for distinguishing certain populations (called ancestry-informative markers or AIMs). Jia et al. [2014] were able to distinguish individuals not only from the continental populations for which the AIMs were selected but also from certain subpopulations as well as individuals that were admixed between two of the populations, using only 35 markers.

Assessing the effectiveness of a classification technique can be challenging. There are several pitfalls that can lead to exaggerated claims of accuracy, for example, a bias arising through testing a classifier with the same data as were used to fit the model. See, for example, the criticism by Brenner [1998] of the analyses of Shriver et al. [1997], as well as the reply [Shriver et al., 1998].

7.5   Inference of phenotype †

Some visual phenotypes, including some facial characteristics (molecular photofitting), hair colour and eye colour, can now be predicted from genotypes with some accuracy. Initial studies linking SNPs to facial morphology were conducted on specific measurements of facial features [Liu et al., 2012]. However, Claes et al. [2014] extended this by using 3D scans in conjunction with SNP typing to allow for prediction of overall fine-scale face shape from a panel of SNPs. The study linked gender, ethnicity and 24 SNPs to changes in face shape, although due to the nature of using 3D scans, it is not possible to link individual SNPs to specific facial feature, but rather to overall changes in face shape that can be quantified. This method allows for the generation of a predicted face shape, somewhat akin to a facial composite drawn from the description given by an eyewitness by a sketch artist. Indeed, the two may be used in conjunction to refine the picture of the target individual.

Various SNPs have been associated with hair colour [Branicki et al., 2011] and eye colour [Walsh et al., 2011]. The HIrisPlex SNP panel [Walsh et al., 2013] for both hair and eye colour has been developed for use in forensic settings. Links between other visual traits and SNPs, such as skin colour [Myles et al., 2007; Beleza et al., 2013], height [Allen et al., 2010] and male baldness [Hillmer et al., 2008], are being investigated but are not yet developed for use in forensic casework.

7.6   Relatedness exercises

7.1   Consider three-locus STR profiles as follows:

	Locus		
Individual	1	2	3
Child c	AB	AA	AB
Mother M	AC	AA	AB
Alleged father Q	BC	AB	AC

Find the overall LR comparing the hypothesis that Q is the father of c with the hypothesis that a man X, unrelated to both Q and M, is the father:

(a) assuming no coancestry (i.e. $F_{ST} = 0$);

(b) assuming $F_{ST} = 5\%$ for Q and X, but no direct relatedness of these two with M.

(c) † assuming $F_{ST} = 5\%$ for each of Q, M and X.

At each locus, assume that $p_A = 10\%$ and $p_B = 5\%$.

7.2   (a) Repeat 1(a) but now assuming that X is the half-brother of Q.

(b) What is the posterior probability that Q is the father of c if there are 12 potential fathers: Q, his half-brother and 10 unrelated men. Use $F_{ST} = 0$, and assume that each man is equally likely a priori to be the father.

7.3   Repeat 1(a) but now assuming that genotypes from a fourth STR locus are also available. At this locus, $G_c = AA$, $G_M = AB$ and $G_Q = CC$. Here assume that every C allele has probability 1 in 1/2000 of mutating to an A in each paternal transmission. (Once again assume that $p_A = 10\%$ and $p_B = 5\%$).

7.4   Using only the three-locus profiles of Q and c given in 1 above, calculate an overall LR comparing

(a) the hypothesis that they are siblings with the alternative that they are unrelated; Use (i) $F_{ST} = 0$ and (ii) $F_{ST} = 2\%$.

(b) † The hypothesis that they are siblings with the alternative that they are father–child; use $F_{ST} = 0$.

8

Low-template DNA profiles

8.1   Background

In recent years, enhancements have been made to DNA profiling protocols to allow profiling of trace and/or degraded DNA. As a result of these enhancements, DNA can be profiled from sources which were previously intractable, such as DNA left from a simple touch, for example, from a bullet casing or door handle. A complex DNA profiling case is described in Lohmueller and Rudin [2012], including discussion of laboratory and analysis issues associated with low-template DNA (LTDNA).

The standard number of PCR cycles used for basic forensic STR profiling is 28. The simplest, and initially the most popular, enhancement is to increase this to between 29 and 34 cycles. Initially, the term low copy number (LCN) profiling was used; however, it was not clear whether this referred to a specific 34-cycle profiling technique; therefore, the term LTDNA profiling is now preferred. 'Template' refers to the available DNA strands for polymerase chain reaction (PCR) and so is equivalent to the amount of DNA initially present in the sample. Use of more than 31 cycles is now uncommon, because of the stochastic effects that co-occur with increased PCR cycles. Alternative enhancements for LTDNA profiling include PCR product purification, longer injection time, higher injection voltage and ionic manipulation of the sample and/or buffer [Butler et al., 2004]; all of these enhancements act to increase the amount of DNA injected into the analyzer.

The peak heights of an electropherogram (epg) give some indication of the amount of DNA used in a profiling run. Figure 8.1 shows the combined panels of an epg obtained from a single contributor at three different DNA templates, approximately 500 pg (Figure 8.1a), 31 pg (Figure 8.1b) and 15 pg (Figure 8.1c). We can see that maximum heterozygote peak heights fall from approximately 3500 RFU at 500 pg

Weight-of-Evidence for Forensic DNA Profiles, Second Edition.
David J. Balding and Christopher D. Steele.
© 2015 John Wiley & Sons, Ltd. Published 2015 by John Wiley & Sons, Ltd.
Companion Website: www.wiley.com/go/balding/weight_of_evidence

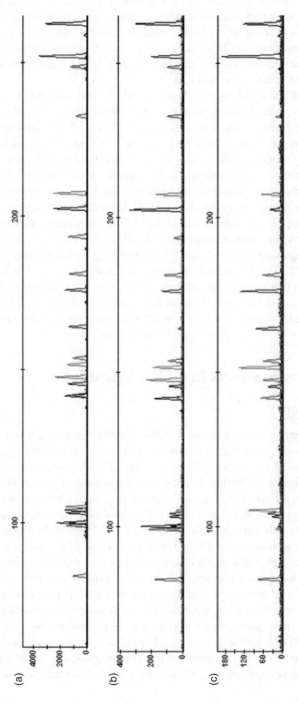

Figure 8.1   The epg from Figure 4.1 is shown again, in part and with all panels combined. Also shown are corresponding results from two profiling runs with low DNA template: 31 pg (b) and 15 pg (c). Note the very different y-axis scales.

to 350 RFU and 190 RFU for the 31 pg and 15 pg panels, respectively. Stochastic artefacts become more prevalent as the DNA template is decreased, which result from a combination of a small number of cells contributing to the sample and the detection of extracellular and degraded DNA in the sample due to the high sensitivity of the technique. The introduced artefacts can include drop-in, drop-out, peak imbalance and exaggerated stutter (see Section 8.2).

Because there is no clear distinction between LTDNA and standard DNA profiling, analyses specialised for LTDNA profiles should return the same results as a standard analysis for profiles from a good-quality DNA sample. For standard profiles, stochastic phenomena are assumed to be negligible, whereas analyses specialised for LTDNA profiles account for these phenomena. A threshold is often proposed, typically between 200 and 300 pg, below which the profile is considered to be LTDNA. However, such a threshold should be determined by the details of the profiling system employed and the sample analysed. Additionally, for mixed DNA samples, quantification techniques cannot estimate the quantity of DNA from each contributor, so one contributor's DNA may be high template, while other contributors (who may include Q) may be low template. While standard DNA profiling protocols typically suggest 500–1000 pg of DNA, they usually perform well with 300 pg of DNA or lower. Figure 8.1 shows that while drop-out does occur, there is still substantial information that can be gained from only 31 pg of non-degraded DNA (Figure 8.1b), which is roughly equivalent to the DNA content of five cells, and indeed 15 pg of DNA (Figure 8.1c) even provides a good deal of usable information.

8.2   Stochastic effects in LTDNA profiles

8.2.1   Drop-out

Drop-out occurs when an allele from a contributor to the crime sample is not reported in the profile obtained from that sample. By comparing Figure 8.1a to Figure 8.1b and c, we can see several instances of drop-out in the LTDNA panels; for example, the allele at approximately 195 bp has dropped out in Figure 8.1c. Peaks that fail to reach a height threshold for detection may be classed as drop-out, as they cannot be reliably distinguished from baseline noise. The detection threshold is specific to laboratories and profiling protocols but is usually between 25 and 75 RFU. In Figure 8.1c, the rightmost peak at approximately 230 bp is within the range of the detection threshold and may or may not be called as allelic depending on the specific threshold used.

DNA degradation from environmental exposure reduces the template available for amplification and so increases the drop-out rate, an effect that tends to increase with DNA fragment length. In good-quality single-contributor profiles, such as Figure 8.1a, a homozygous genotype at a locus is represented by a single peak. However, in a LTDNA profile, when a single peak is detected at a locus, the possibility that a second allele has dropped out must be recognised by the interpreter, for example, the peak at approximately 80 bp in Figure 8.1c. A threshold is set by some laboratories, typically around 300 RFU, above which a single peak is interpreted as homozygote. Such a threshold is used for the UK National DNA Database to call a single peak at allelic position A as either AA or AF, where F will match any allele in subsequent database searches. A method to determine this threshold based on specific profiling protocol calibration data was proposed by Puch-Solis et al. [2011].

8.2.2   Drop-in

Drop-in refers to a crime scene profile (CSP) allele that has not come from any of the assumed contributors to a DNA sample. A drop-in allele must come from someone, but if the DNA from that individual is very low template and/or degraded, it may be appropriate to treat it as a sporadic event rather than as originating from an extra contributor. The appearance of sporadic alleles in DNA-negative controls confirms the existence of drop-in within LTDNA work, possibly arising from airborne DNA fragments generated by previous analyses in the same laboratory, motivating the term 'drop-in'. However, drop-in alleles could arise from environmental exposure to degraded DNA at the crime scene. Only laboratory-based contamination is considered to be drop-in by some authors; however, verification of the source of a drop-in allele is not usually possible. There are similar issues with use of the term 'contamination': it often refers to any foreign DNA introduced into the sample after recovery from the crime scene. While this distinction may be important at trial, the epg provides little information on the times of deposition of different sources to the DNA sample. As a result, this term is also used more broadly to refer to any DNA originating from anywhere other than the persons of interest for the investigation, which will include drop-in alleles. 'Gross contamination' refers to contamination in which a full complement of an individual's DNA is observed in the crime sample, whereas 'environmental contamination' refers to smaller amounts of DNA, such as drop-in.

8.2.3   Peak imbalance

Peak imbalance, sometimes referred to as heterozygote (im)balance, concerns the heights of the two peaks from an individual at a heterozygote locus. These peaks are expected to be of similar heights, because they come from the same individual, and, therefore, have the same DNA template (see Figure 8.1a). However, peak heights are more variable for LTDNA samples, for example, the heterozygote peak heights at approximately 150 bp differ by a factor of about two on Figure 8.1c. Extreme peak imbalance can cause allelic drop-out. It used to be thought that the high variability of LTDNA peak heights meant that they conveyed little useable information about DNA template, and so the peak height information was not used other than to designate the presence/absence of alleles. However, statistical models incorporating peak heights have been introduced (Section 8.3.7), allowing information on the heterozygote/homozygote status of contributors, and how many contributors have the allele, to be inferred from the peak heights.

8.2.4 Stutter

As we discussed in Sec 4.1.1.2, stutter peaks are non-allelic peaks in an epg that originate due to errors in DNA copying during PCR. Usually, a stutter peak is one repeat unit shorter than an allelic peak, caused by the omission of one repeat unit of the DNA motif in the copying process. Omission of two repeat units (termed 'double stutter') or addition of a repeat unit ('over stutter') also occur. Stutter occurs in standard DNA profiles (Figure 8.1a) but can be exaggerated in some LTDNA protocols. When there is a minor contributor to the sample, stutter can be problematic because a stutter peak from a major contributor may be hard to distinguish from an allelic peak of the minor contributor. A threshold is often used, so that a peak is treated as stutter if it is below the threshold fraction (often 15%) of the 'parent' allele peak height. A single 'hard' threshold is clearly unsatisfactory, and in Chapter 9, we discuss an improvement in which peaks in stutter positions lying between two thresholds are designated as 'uncertain' and peak height models can explicitly model stutter and allelic peak heights (Section 8.3.7).

8.3   Computing likelihoods

8.3.1   Single contributor allowing for drop-out

In the case that CSP = A and G_Q = AB (Table 8.1, middle row), the likelihood under H_p (L_p^1) is the probability that allele B from Q has dropped out (D), while allele A has not $(1 - D)$. In the defence likelihood (L_d^1), to explain the CSP, X must be either AA or AF, where F is any allele other than A and has dropped out. These lead to the first and second terms of L_d^1, requiring zero and one drop-out, respectively. The defence and prosecution values for D need not be the same, but similar values are often supported if the numbers of contributors under H_d and H_p are equal, and a single D common to H_d and H_p is often assumed for LR calculation demonstrations [Gill et al., 2007].

Table 8.1   Likelihoods under defence (L_d^1) and prosecution (L_p^1) hypotheses for single-contributor profiles when drop-out is required to explain the CSP under H_p. We assume that the epg has been interpreted to designate presence/absence of alleles, with no further use of peak heights. Drop-out events are assumed to be distributed as independent Bernoulli trials, with probability of success D (D_2 for homozygotes). Ø denotes no observed alleles and P_{hom} is the fraction of the population that has a homozygous genotype.

| | | L_p^1 | |
CSP	L_d^1	G_Q = AB	G_Q = AA
AB	$2p_A p_B (1 - D)^2$	$(1 - D)^2$	0
A	$p_A^2 (1 - D_2) + 2p_A(1 - p_A)D(1 - D)$	$D(1 - D)$	$(1 - D_2)$
Ø	$P_{hom} D_2 + (1 - P_{hom})D^2$	D^2	D_2

Adapted from Table 1 of Steele and Balding [2014a].

In the case that CSP = A and $\mathcal{G}_Q$ = AA, no drop-out events are required to explain the CSP, and if D is assumed to be zero, then the LR comparing H_d with H_p is p_A^2, which is the LR for good-quality profiles. If CSP = AB and $\mathcal{G}_Q$ = AB, then LR = $2p_A p_B$, once again the same as that from a good-quality sample, but now this holds irrespective of the values of D and D_2. Note that if $\mathcal{G}_Q$ = AA, the CSP cannot be explained under H_p (as we assume no drop-in), and hence, $L_p = 0$.

8.3.2   Profiled contributors not subject to drop-out

'Mixed' profiles (with DNA from more than one individual) are frequently encountered in LTDNA work. Often the sample contains a high DNA template from a profiled contributor, K (who may be a victim, for instance), in addition to a low-template contributor of interest, who may be the offender, alleged to be Q. If K and Q have no shared alleles at a locus, the presence of alleles from K has no effect on the likelihood (Table 8.1). If an allele is shared between the two, one or both alleles of Q are 'masked' by alleles of K. Single-locus likelihoods for some examples of masking are given in Table 8.2, under the following hypotheses:

$$H_d^2 : \; X + K \qquad \text{and} \qquad H_p^2 : \; Q + K.$$

When CSP = ABC, $\mathcal{G}_Q$ = AB and $\mathcal{G}_K$ = BC, the LR comparing H_d^2 and H_p^2 can be written as

$$\text{LR} = \frac{L_d^2}{L_p^2} = \frac{\sum_{g \in \Gamma} p_g P(\text{CSP} = \text{ABC} | \mathcal{G}_X = g, \mathcal{G}_K = BC)}{P(\text{CSP} = \text{ABC} | \mathcal{G}_Q = AB, \mathcal{G}_K = BC)}, \tag{8.1}$$

Table 8.2   Likelihoods under hypotheses H_d^2 (L_d^2) and H_p^2 (L_p^2) for a two-contributor CSP at a single locus, when K is a profiled contributor not subject to drop-out.

CSP	$\mathcal{G}_K$	L_d^2	L_p^2	
			$\mathcal{G}_Q$ = AB	$\mathcal{G}_Q$ = AA
ABC	BC	$p_A^2(1 - D_2) + 2p_A(p_B + p_C)(1 - D) +$ $2p_A(1 - p_A - p_B - p_C)D(1 - D)$	$1 - D$	$1 - D_2$
AB	BB	$p_A^2(1 - D_2) + 2p_A p_B(1 - D) + 2p_A$ $(1 - p_A - p_B)D(1 - D)$	$1 - D$	$1 - D_2$
BC	BC	$(p_B + p_C)^2 + 2(p_B + p_C)(1 - p_B - p_C)D +$ $P_{het}D^2 + P_{\text{hom}} D_2$	D	D_2
B	BB	$p_B^2 + 2p_B(1 - p_B)D + P_{het}D^2 + P_{\text{hom}} D_2$	D	D_2
AB	AB	$(p_A + p_B)^2 + 2(p_A + p_B)(1 - p_A - p_B)D +$ $P_{het}D^2 + P_{\text{hom}} D_2$	1	1

Here, P_{het} and P_{hom} denote the population fractions of heterozygous and homozygous genotypes that do not include any of the CSP alleles.
Adapted from Table 2 of Steele and Balding [2014a].

where p_g indicates the population frequency of genotype g and Γ indicates the set of possible genotypes. The first row of Table 8.2 gives expressions for the numerator and denominator. L_d^2 is derived by summing the product of the population genotype frequency and masking/drop-out/non-drop-out terms similar to those in L_p^2, over each possible genotype for X. L_p^2 is derived by considering whether or not each allele of Q has dropped out, or whether this cannot be determined due to masking by an allele of K.

This model was extended by Balding [2013] to allow a third, uncertain, category for each allelic position in the epg, in addition to present or absent. The uncertain category is useful for peaks that may be attributable to stutter or other artefact or for peaks that have a borderline peak height. Uncertain alleles are handled in the same way as alleles masked by K; the drop-out term is one, as it is unknown whether or not the allele has dropped out.

8.3.3   Modelling drop-in

Drop-in is often modelled as an independent Bernoulli event, with the allelic classification of the drop-in being proportional to p, but excluding all alleles of the hypothesised contributors. When unprofiled contributors are considered, for example, X under H_d, an allele seen in the CSP can either be included in the genotype of X or be treated as drop-in, and this designation will be different for each term of the sum. The probability of a drop-in event at a given locus thus depends on the relative frequencies of the hypothesised contributor's alleles, as potential drop-in events that are masked by a contributor's allele will not count as drop-in, implying a varying probability of drop-in in the terms of the summation. However, this complexity may be ignored in practice.

It is possible to avoid modelling drop-in, by instead including an additional unprofiled contributor of very LTDNA. The drop-in model may, however, reduce the computational effort and provide an acceptable approximation. In work not shown here, we simulated 500 two-replicate CSPs with no drop-out for the first contributor, and five levels of drop-out for the second contributor. The LR was evaluated for each CSP assuming both $Q/X + U$ and $Q/X +$ drop-in as the hypotheses. We found that for realistic levels of drop-in, and up to 10 alleles not attributable to Q, the two hypothesis pairs led to very similar LRs.

If unlimited drop-out and drop-in are allowed, any hypothesis for the contributors has non-zero probability for any CSP, although the probability may be small if many drop-ins and drop-outs are required to explain the CSP. Therefore, it may be reasonable in practice to place a limit on the number of drop-ins permitted at a locus.

8.3.4   Multi-dose drop-out and degradation

A CSP with no observed alleles has LR $\neq 1$ and is, therefore, informative (see bottom row of Table 8.1). This is because the homozygote drop-out probability, D_2, should be less than D^2, the probability that two heterozygote alleles have both dropped out. Based on limited data, Balding and Buckleton [2009] used $D_2 = D^2/2$ in their numerical computations, but this approximation means that D_2 can never exceed 0.5, which is unsatisfactory [Tvedebrink et al., 2009]. These authors instead suggested modelling drop-out probabilities as a function of DNA 'dose' for which average peak height served as a proxy. When multiple contributors are subject to drop-out, they may contribute different amounts of DNA for a shared allele, necessitating a general formula.

From Tvedebrink et al. [2009], $D(k)$, the probability of drop-out for dose k of DNA can be written as

$$\frac{D(k)}{1 - D(k)} = (\alpha_s k)^\beta, \tag{8.2}$$

where s indicates the locus. If k is defined so that the probability of heterozygote drop-out is $D(1)$, then the probability of homozygote drop-out is $D(2)$. When k is large

$$\frac{D(2k)}{D(k)^2} \approx \left(\frac{2}{\alpha_s k}\right)^\beta > 1,$$

meaning drop-out of a homozygote allele can be more likely than drop-out of two heterozygote alleles, but this only arises for very high drop-out probabilities and is unimportant in practice.

Buckleton et al. [2014] removed the dependency of the drop-out rate on the locus being considered. They show that the removal of locus dependency coefficients results in a more robust drop-out model when using empirically estimated drop-out coefficients, as a coefficient for each locus can lead to overfitting to the training data. However, this is not relevant to models that maximise over the model parameters, as then locus-specific coefficients are fit to the CSP without the need for training data.

In Puch-Solis et al. [2013a] and Cowell et al. [2015], the drop-out probability corresponds to the lower tail of a gamma distribution. Cowell et al. [2015] noted that for typical parameter values, their definition implies a slower increase in the drop-out rate as the DNA dose increases than that under the model (8.2). Threshold-based definitions necessarily satisfy $D_2 < D^2$; however, no goodness-of-fit comparisons have been made between these models.

The degradation rate of DNA depends on humidity, temperature and environmental exposure. The peak heights in an epg tend to decline approximately exponentially with fragment length, in part due to degradation [Bright et al., 2013a]. A geometric model for the effective DNA template at a given allele fragment length was proposed by Tvedebrink et al. [2012], where longer fragments have lower effective allele dose. Repeat number is a poor proxy for fragment length, because STRs include flanking regions on either side of the core repeat region used for primer binding; actual fragment lengths for many DNA profiling systems can be obtained from the STR-base website (http://www.cstl.nist.gov/strbase/). For some aspects of LTDNA analysis (e.g. stutter analysis), the longest uninterrupted sequence (LUS) of the STR may be more applicable than the fragment length or repeat number [Kelly et al., 2013].

8.3.5   Additional contributors subject to drop-out

Once a multi-dose drop-out model has been specified, likelihoods can be calculated for any number of profiled/unprofiled contributors, each of whom may or may not be subject to drop-out. In the context of (8.1), if we assume an unprofiled contributor, U, who is subject to drop-out, rather than K, then both the numerator and denominator have to be summed over all possible genotypes for U, as usual multiplying each term by the genotype probability:

$$
\text{LR} = \frac{\sum_{g1,g2 \in \Gamma} p_{g1} p_{g2} P(\text{CSP=ABC} | \mathcal{G}_X = g1, \mathcal{G}_U = g2)}{\sum_{g \in \Gamma} p_g P(\text{CSP=ABC} | \mathcal{G}_Q = AB, \mathcal{G}_U = g)}.
\tag{8.3}
$$

Information about the relative DNA template from each hypothesised contributor is necessary to compute the likelihood, because more than one contributor is subject to drop-out. This information can be derived from peak heights assumed to have originated from a single contributor [Tvedebrink et al., 2009], or DNA templates can be treated as unknown parameters, and eliminated through maximisation or integration of the likelihood [Cowell et al., 2015; Balding, 2013]. Taylor [2014] demonstrated that as the amount of DNA from Q tends towards zero, so does the log(LR); with little DNA from Q there is little information available in the CSP to establish that Q is a contributor.

8.3.6   Replicates

Replication is valuable in accounting for variability in a measurement and, therefore, seems desirable to handle stochastic aspects of LTDNA profiles. Signals that occur in multiple replicates can then be distinguished from artefacts that occur only in a single replicate. Between two and four replicate profiling runs are routinely performed in the United Kingdom in LTDNA work. Mitchell et al. (2012) reported using one or two 28-cycle replicates if the DNA template is ≥ 300 and three 31 cycle replicates otherwise. Note that in LTDNA work, the term 'replicate' is often not strictly correct, because some profiling parameters are varied in different runs, but we follow the custom of still referring to these as replicates.

Initially, 'consensus' profiles were recommended [Gill et al., 2000], which manually combined information from different replicates into a single CSP, for example, by including only replicated allelic signals. Optimal approaches for combining multiple replicates were developed by Benschop et al. [2011], who reported that the tools needed for a statistical analysis of LTDNA mixtures were not available at the time. They are now available, rendering consensus profiles largely obsolete for the purpose of evaluating weight of evidence.

Replication has both proponents and opponents; on one hand, replication is seen as a cornerstone of the scientific method and should be used wherever feasible. Pfeifer et al. [2012] said 'Since in many cases only partial profiles can be obtained from LT samples, and the truncated profiles are usually difficult to interpret, replication and optimization of DNA typing are generally considered to overcome these constrictions'. On the other hand, replication will split what is often an already minute sample [Grisedale and van Daal, 2012], and instead, it would be preferable to use as much as possible of the available DNA, giving the best possible single-run profile. Parameter estimation is still possible from a single profiling run, which is achieved by utilising information available across loci, as different loci are, to some extent, profiling replicates subject to the same experimental conditions.

Expressions for LR calculations including multiple replicates and allowing for both drop-out and drop-in were given by Curran et al. [2005]. Replicates are typically assumed to be independent, conditional on unprofiled contributor genotypes and other parameters such as those measuring DNA dose from each contributor. The problem lies in specifying the parameters conditional on which the independence assumption is reasonable. Under the prosecution hypothesis of $Q + U$, the multi-replicate likelihood can be expressed in the following form:

$$L = \sum_{g \in \Gamma} p_g \prod_r P(\text{CSP}_r | Q, \mathcal{G}_U = g, \psi), \tag{8.4}$$

where CSP_r indicates the set of all alleles observed in the rth replicate and ψ is a set of parameters.

8.3.7   Using peak heights

Hitherto we have assumed that epg interpretation results only in decisions about the presence/absence of alleles (or possibly an uncertain designation). However, peak heights hold information about allele dose, which is useful for mixed profiles, as can be seen in Figure 8.2. Locus *D16S539* shows four approximately equal-height peaks, indicating two contributors of approximately the same DNA template so that all six possibilities for the genotypes of the two contributors are equally supported and the peak heights convey no information beyond the fact that four alleles are present. In contrast, locus *D10S1248* shows three peaks, with the peak at allele 14 being approximately twice the height of those at alleles 13 and 15. This combination of peak heights suggests that the two contributors are 13,14 and 14,15 or 14,14 and 13,15; these genotype combinations are supported much more than genotypes that include two copies of the 13 or the 15 alleles and only one copy of the 14. For this locus, peak heights convey considerable information about the underlying genotypes beyond the mere presence/absence of alleles.

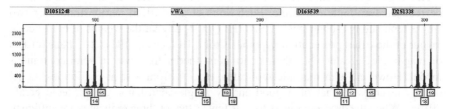

Figure 8.2   The Blue panel of an epg from standard DNA profiling of a mixture with equal DNA template from two contributors.

The peak height variability inherent in LTDNA work means that categorical allele count inferences are not usually possible, but useful inferences about contributor genotypes through DNA template can be gained through a suitable statistical model predicting DNA template based on peak heights [Perlin and Szabady, 2001; Perlin and Sinelnikov, 2009]. DNA profiling systems also return peak areas [Evett et al., 1998; Gill et al., 1998]; however, peak areas and heights are highly correlated [Tvedebrink et al., 2010], with heights usually preferred over areas. A discussion of some of the defence challenges to peak height data and models is given in Taylor [2014].

Models that use peak heights (termed continuous models) give likelihoods with the same structure as shown previously, for example, (8.4); however, the peak height of each allele at every locus is now represented in the CSP, rather than presence/absence/uncertain. The peak heights are often assumed to be gamma distributed [Cowell et al., 2007], with the mean corresponding to DNA template, and departures from the mean over the entire profile used to estimate the variance. An alternative to the gamma distribution is the lognormal distribution. Stutter peak height can also be modelled as a gamma distribution [Puch-Solis et al., 2013a], with mean calculated as a fraction of the parent allele mean. A threshold of detection is sometimes used, below which any peak is ignored; this is to deal with baseline peaks over the entire epg and means that models that use this threshold are not fully continuous: above the threshold the peak height distribution is continuous, but the overall distribution has an atom of probability mass which corresponds to drop-out (meaning below-threshold peak height).

The pattern of variability for LTDNA profile peak heights can depend sensitively on the specifics of the DNA profiling protocol used, which may restrict their effectiveness [Gill and Haned, 2013]. Some continuous models use only the CSP data to estimate model parameters [Perlin et al., 2011a; Cowell et al., 2015], while others need laboratory calibration data generated under conditions matching those from when the CSP under evaluation was generated [Taylor et al., 2013; Puch-Solis et al., 2013a]. Therefore, there is a trade-off between additional statistical efficiency and generality of the model in question. Modelling assumptions may have impacts on continuous models that do not rely on calibration data, so even these models may behave differently depending on the details of the protocol used. Semi-continuous models (those which use allele presence/absence but allow for drop-out and drop-in continuously) are subject to similar problems; however, the simpler modelling assumptions and data act to reduce these concerns. The robustness of LTDNA models to varied profiling platforms and protocols seems not to have been extensively investigated.

The largest gain in statistical efficiency from use of peak height information will be realised for single-run CSPs. Puch-Solis et al. [2013a] demonstrated for two single-replicate two-contributor CSPs (H: $Q/X + U$) that the weight of evidence in favour of the (true) prosecution hypothesis for a 10-locus profile is 2–3 bans (orders of magnitude) greater with their continuous model than with a semi-continuous model. However, with multiple replicates, the LR from a semi-continuous model often converges to that of a good-quality single-contributor CSP [Steele et al., 2014a]. When the drop-out rate is low, as few as two or three replicates are sufficient to approach this optimal LR.

Throughout this book, we have approached the evaluation of weight of evidence from the viewpoint of a competition between prosecution and defence hypotheses, H_p and H_d, and treated computation of the likelihoods, including elimination of nuisance parameters, separately under H_p and H_d. Often, however, H_p is a special case of H_d obtained by replacing an unprofiled contributor X with a queried contributor Q. In that case, it may only be necessary to perform computations under H_d, from which results for H_p can be derived. This is most apparent for Bayesian MCMC software [Perlin et al., 2011a], when the state of the Markov chain varies over possible genotypes for the unprofiled contributors, and the LR can be expressed in terms of the fraction over the MCMC run of iterations in which the state visited is consistent with H_p. Perlin et al. [2011a] claimed an advantage that this approach can give a demonstrable avoidance of pro-prosecution bias because the MCMC can be run without knowledge of H_p. However, in practice, knowledge of H_p favours the defence through the inclusion of an F_{ST} adjustment for the alleles of Q. Further, if the evidence implicating Q is relatively weak, there may be few visits of the chain to states consistent with H_p, and so nuisance parameters conditional on H_p may be poorly estimated, leading to an imprecise LR estimate. For example, if the match probability for the profile of Q is of the order of 10^{-18} and the likelihood for a complex CSP to which Q is a contributor is about 10^{-8}, this corresponds to on average one visit to H_p for every 10 billion MCMC iterations, which will be difficult to approximate empirically. In practice, it may be more efficient to run two separate MCMCs, conditioning on H_d in one and H_p in the other.

8.4   Quality of results

Validation is possible for many laboratory procedures by ensuring that measured quantities reliably lie within an acceptable range of error from the known standard. However, no such true value is available for LR calculations, and therefore, this kind of direct validation is not possible. The LR measures our uncertainty about a binary event, and it depends on modelling assumptions. In LTDNA work, the samples are often degraded and minuscule, making verification of these assumptions difficult. Open-source software facilitates external scrutiny, thus generating suggestions for improvements or bug reports.

The match probability for a fully represented single contributor acts as a theoretical lower bound on the LR of a CSP when that individual is Q [Cowell et al. 2015]. As a result Cowell et al. [2015] proposed the ratio of the LR to the match probability as a measure of evidential efficiency. Even if individual profiling runs are noisy, with multiple runs, the LR will often converge to the match probability and, hence, to the maximum efficiency [Steele et al., 2014a]. However, the LR may not approach the match probability if, in a multi-contributor scenario, all contributors provide approximately equal templates to the sample, as the different contributors cannot be distinguished through differential drop-out rates and/or differential expected peak height contributions. The existence of the match probability bound itself can act as a tool to validate programs for evaluation of LTDNA LRs from multi-replicate CSPs [Steele et al., 2014a].

False H_p performance tests are advocated by Gill and Haned [2013] and Dørum et al. [2014]. These are useful both to investigate the behaviour of various models and to check that a specific program correctly implements the specified model. Some LTDNA analysis programs implement tests in which the genotype of Q is replaced by a random genotype generated under a standard population genetics model. This can lead to an inference of the percentage of random Q LRs that the true LR exceeds, which should be large if the model is reasonable, as the majority of randomly generated Qs will not closely match the real Q and will, therefore, explain the CSP poorly and return a large LR. The computational cost for other models is likely to be too high to routinely perform this type of performance test.

False 'inclusion' is always possible; what is salient is how unlikely it is, which the LR conveys. Balding [2013] demonstrated that a random Q often leaves many CSP alleles unexplained, necessitating an extra contributor under H_p than is strictly necessary under H_d. Consequently, an extra contributor may be invoked under H_d as well; however, this can be computationally expensive, while potentially hindering the performance of some models. Inclusion of an extra contributor under H_d will be slightly beneficial to the defence if the model used can adequately discern the relative template from each hypothesised contributor. While these performance tests cannot make any inference about the evidential strength in a particular case, they can give an idea of the average performance of a model.

Due to the shifting landscape of LTDNA analysis software, few systematic comparisons of the various programs have been published. Now that the field is starting to mature, it is important that such comparisons are performed. The currently available results suggest that the same hypothesis pair does produce different results depending on the program used. The principal differentiation is between semi-continuous and continuous algorithms and stems from the utilisation of peak height information in the latter. The weight of evidence can be up to several bans difference between the two for single-replicate samples; however, the difference is usually less pronounced with multiple replicates. Smaller differences are seen between programs within each class of model, which result from distinct modelling assumptions and/or methods for the elimination of nuisance parameters; however, these differences are small compared to the typical LR range in practice. Thus, extreme precision of likelihood calculations is futile. Relative differences of up to a factor of two should be seen as inconsequential, as the effect of different (and valid) modelling assumptions often exceed this level. In fact, typical F_{ST} and sampling adjustments can alter the LR by one to two bans, intended to favour the defence.

9

Introduction to `likeLTD` †

`likeLTD` (likelihoods for low-template DNA profiles) is an R package for computing likelihoods for DNA profiles. It is particularly suited for low-template and/or degraded DNA when alleles from some contributors may be subject to drop-out. It can handle multiple profiled possible contributors and up to two unprofiled contributors, in addition to Q/X. The package also provides input files for an example analysis (the 'Hammer Case' described below).

Here we show how to install and run `likeLTD` using an illustrative example. We also describe the model underlying `likeLTD`, for example, explaining the 'uncertain' category for allele designations and the drop-out and degradation models. We present results of running `likeLTD` on a range of single contributor and mixed DNA profiles subject to modifications, such as introduction of artificial drop-out and drop-in. Some of the material used here, and some other analyses, are published in Balding [2013].

In the results demonstrated here, we have used Version 5.4.0 of `likeLTD`, with a standard allele frequency database of around 200 UK Caucasians, $F_{ST} = 0.02$ and a sampling adjustment adj $= 1$.

9.1 Installation and example R script

Both installing `likeLTD` (only needs doing once on any computer) and loading it (once per R session) are very simple.

```
install.packages("likeLTD")
require(likeLTD)
```

The `install.packages` command may generate a request for you to choose a site from which to download the package. Choose any site near you.

Weight-of-Evidence for Forensic DNA Profiles, Second Edition.
David J. Balding and Christopher D. Steele.
© 2015 John Wiley & Sons, Ltd. Published 2015 by John Wiley & Sons, Ltd.
Companion Website: www.wiley.com/go/balding/weight_of_evidence

The example analysis that comes with likeLTD is called the Hammer Case. The DNA profiles are from Table 2 of Gill et al. [2007], who introduced the software LoComatioN which in some respects is similar to likeLTD and also to LRmix Studio (http://lrmixstudio.org/). The crime scene profile (CSP) consists of two profiling runs at each of 10 loci. These, and reference profiles from a queried contributor Q and two victims ($K1$ and $K2$), are available in input files hammer-CSP.csv and hammer-reference.csv. likeLTD allows 'uncertain' allele calls, but this designation was not used by Gill et al. [2007], and so there are no alleles labelled as uncertain in this example.

There is a total of six alleles, all of them are unreplicated, that are not attributable to any of Q, $K1$ or $K2$. No more than two of these occur at any one locus. This suggests a comparison of the following two hypotheses for the contributors of DNA to the sample:

$$H_p : \quad Q + K1 + K2 + U1$$

$$H_d : \quad X + K1 + K2 + U1.$$

9.1.1 Input

We now show how to calculate likelihoods under H_p and H_d using likeLTD. The first command below finds out where your system has stored the Hammer Case files and saves that location in datapath. For your own analyses, you will need to create your own CSP and reference files, in the same format as hammer-CSP.csv and hammer-reference.csv. It is usually most convenient to create these files in a specific directory and then set that to be the working directory for R using the command setwd() or using the R menu option (its location varies across operating systems). For example, if your case files are in the directory C:/Users/JoeBloggs/Cases/JoeBloggs1 then you enter the command setwd("C:/Users/JoeBloggs/Cases/JoeBloggs1"). In that case, you can set datapath = "." in place of the first command below. A default allele frequency database file is provided with likeLTD. To use your own database file instead (must be in same format), set databaseFile to the filename, including path if not in the working directory. If you wish to choose a different individual to be Q, or to add or omit one of the other profiled contributors (a K, in the notation used here, standing for 'known'), then you must create a new reference file.

```
datapath = file.path(system.file("extdata",
    package="likeLTD"),"hammer")

# File paths and case name for allele report
admin = pack.admin.input(
            cspFile = file.path(datapath, 'hammer-CSP.csv'),
            refFile = file.path(datapath,
                                'hammer-reference.csv'),
            caseName = "hammer"
            )
```

Values are required for `cspFile` and `refFile`, but `caseName` can be omitted from the above in which case it defaults to `"dummy"`. Two other arguments have been omitted so that their default values will be used: `databaseFile = NULL` and `outputPath = getwd()`.

9.1.2 Allele report

```
# Next we generate an allele report
allele.report(admin)
```

The allele report is a .doc file that will be created in the current working directory (set `outputPath` to specify a different directory). It summarises the input data, highlights rare alleles and suggests values for key parameters, in particular, the number of unprofiled contributors required to explain the observed CSP under H_p and whether modelling drop-in is appropriate. Here, it indicates that if drop-in is modelled, only one unknown contributor is sufficient under H_p to explain the observed alleles not attributable to Q/X or $K1$ or $K2$. While it is never possible to specify an upper bound on the number of known contributors, specifying more than the minimum required usually has negligible impact on the resulting likelihood ratio (LR). Cowell et al. [2015] illustrated this with an example in which a $\log_{10}(LR)$ of 14.09 with three contributors barely changes as the number of contributors increases, reaching 14.04 with eight contributors.

9.1.3 Arguments and optimisation

Based on the allele report, we specify the required hypotheses by setting a list of arguments containing the following items:

nUnknowns. The number of unknown contributors (0, 1 or 2) under H_p. likeLTD automatically adds an additional unknown contributor (X) under H_d, who replaces Q in H_p.

doDropin. Whether to model drop-in or not (logical: TRUE or FALSE).

ethnic. The database population most appropriate for Q. The default database comes with 'EA1','EA3' and 'EA4', corresponding to UK residents of Caucasian, Afro-Caribbean and South Asian origin, respectively. If you use your own allele frequency database, you will choose your own population labels (required even if there is only one population).

adj. Sampling adjustment (scalar).

fst. F_{ST} adjustment for coancestry of Q and X (scalar).

relatedness. Relatedness coefficients for Q and X (vector of length two): the probabilities that they have one and two alleles identical by descent from recent common ancestors (e.g. parents or grandparents). The setting used below is for Q and X unrelated; for siblings, use `relatedness = c(0.5, 0.25)`. Version 5.4.0 of likeLTD does not take into account linkage between markers, which will lead to a slight overstatement of the strength of evidence. The user may prefer to omit one of each pair of syntenic loci, which will instead lead to understatement of the strength of evidence. An approximate linkage correction is planned for the next version.

```
# Enter arguments
args = list(
        nUnknowns = 1,
        doDropin = FALSE,
        ethnic = "EA1",
        adj = 1,
        fst = 0.02,
        relatedness = c(0,0)
        )

# Create hypotheses
hypP = do.call(prosecution.hypothesis, append(admin,args))
hypD = do.call(defence.hypothesis, append(admin,args))

# Get parameters for optimisation
paramsP = optimisation.params(hypP)
paramsD = optimisation.params(hypD)

# Run optimisation
results = evaluate(paramsP, paramsD)
```

The entries in `args` shown above are defaults, except for nUnknowns which defaults to 0. The function `do.call` calls the function given in its first argument. Both `prosecution.hypothesis` and `defence.hypothesis` are functions defined within `likeLTD`, which generate the necessary objects for H_p and H_d, respectively. The `evaluate` function is likewise defined within `likeLTD` and is a wrapper function for the `DEoptim` function that performs optimisation (see Section 9.2.3), providing improved optimisation and a progress bar displaying current weight of evidence (WoE). The progress bar can be disabled by setting the argument `progBar = FALSE`, which is necessary if you do not have graphical capabilities, for example, running from command line on a server. The `evaluate` function now splits the convergence into a number of steps, with each subsequent step having more stringent convergence tolerance and an increased crossover rate (a parameter for `DEoptim`); the combination of these two behaviours means that the parameter space is searched extensively to start with and gradually anneals to a more intensive local search towards the end. The number of steps to run is determined by the difficulty of converging the first step. Interim results after each step are available by setting the argument `interim = TRUE`, which writes the most recent results to `Interim.csv`, in the current working directory. The function `optimisation.params` sets the parameter values needed for DEoptimLoop. These values can be altered if required, but the default settings should be adequate for most analyses.

The object returned by `evaluate` is a list of three elements: `Pros`, `Def` and `WoE`. Both `Pros` and `Def` have the same structure as the object returned by `DEoptim` (see `help(DEoptim)`), with each corresponding to the prosecution and defence results, respectively. `WoE` gives the WoE for each step run by `evaluate` in bans, the final WoE can be obtained through the command `results$WoE[length(results$WoE)]`.

9.1.4 Output report

```
# Generate output report
output.report(hypP,hypD,results)
```

The results are given in the output file `hammer-Evaluation-Report-1.doc` (the numbering of the filename increments automatically, or a custom filename may be specified with `file="fileName.doc"`) which again summarises the input data, similar to the allele report, and also states the hypotheses compared and gives single-locus and overall LRs. Logarithms (base 10) are also given for the LRs, which are expressed as support for H_p relative to H_d, which as we explained in Section 3.1 is a widely used convention but the opposite of what we have been using in this book up to the present Chapter. We use WoE to denote weight of evidence, in bans (= powers of 10). For the Hammer case, the overall WoE was found to be 10.9 bans. The WoE is > 0 (favours H_p over H_d) at every locus except D18 (WoE = -0.5 bans). The most incriminating locus is D19 (WoE = 2.6 bans), where the two alleles of Q are rare, replicated in the CSP and not shared with either $K1$ or $K2$.

Estimates of the drop-out rate for each contributor subject to drop-out, and each profiling run, are also given in the output file. These estimates are often not precise, particularly under H_d where there are two unprofiled contributors, but this is not important for assessing the WoE against Q. Note that under H_d, X and $U1$ are indistinguishable; we usually take X to be the one with drop-out rates most similar to Q, but the labelling in the output file is arbitrary. The single-allele drop-out rates for Q/X are estimated at 11%/12% and 0%/1% in the two replicates. The drop-out rate estimates for $K1$ under H_p/H_d are 46%/44% for replicate a and 3%/5% for replicate b. The degradation value for $K1$ (γ_{K1}) is about 0.8% under both hypotheses, while γ_Q and γ_X are both about 0.4%, indicating an increase in drop-out rate with fragment length, an effect of degradation, particularly for $K1$.

Every CSP allele attributable to $K2$ could also come from $K1$ or Q, and so under H_p, there is no evidence for DNA from $K2$. However, under H_d, the DNA of Q is not present, leaving three CSP alleles that can be attributed to $K2$ but not to $K1$. Nevertheless, `likeLTD` estimates 100% drop-out of the alleles of $K2$ in both replicates and under both hypotheses. This is because the three alleles attributable to $K2$ under H_d are all replicated, whereas seven other alleles of $K2$ do not appear at all, indicating very high drop-out, and so `likeLTD` finds that attribution of the three alleles to $K2$ is unlikely. Although we cannot exclude $K2$ from contributing any DNA to the sample, these results indicate that including $K2$ in the analysis brings no explanatory power and so has negligible impact on the WoE implicating Q as a contributor. It is not necessary to exclude $K2$ from the analysis, because likeLTD has automatically done this by estimating drop-out at 100%, but there may be a slight improvement in the likelihood optimisation in running the analysis again without $K2$, due to fewer nuisance parameters to be estimated.

9.1.5   Genotype probabilities

likeLTD can provide a list of the most probable genotypes at each locus for each unprofiled contributor, using function get.likely.genotypes. By default, only single-locus genotypes with probability > 0.1 are returned; this can be altered using the argument prob. The most probable whole-profile genotype, and its probability are also returned (Figure 9.1). The genotypes and their probabilities add nothing to the assessment of WoE against an alleged contributor of DNA but can be useful for searches in a database.

```
# Get the most likely single-contributor genotypes
gens = get.likely.genotypes(hypD,paramsD,results$Def)
```

The returned list object is organised into a series of levels, as shown in Figure 9.1. It may also be desirable to obtain the probabilities of joint genotypes, rather than genotypes of single contributors. In this case, the argument joint can be handed to get.likely.genotypes, and if set to TRUE, the joint genotypes and probabilities will be returned rather than the single-contributor genotypes and probabilities. For joint genotypes, the default probability threshold for single-locus genotypes is 5%. This value can be altered for the single-contributor or joint cases by setting prob.

Note some of the locus-specific genotypes used to construct the whole-profile genotype may have smaller probabilities than the threshold and will, therefore, not be displayed in the locus-specific results.

```
# Return joint genotypes and probabilities
gens = get.likely.genotypes(hypD,paramsD,results$Def,
                            joint=TRUE)
```

```
# Return joint genotypes with per-locus probability
  # greater than 3%
gens = get.likely.genotypes(hypD,paramsD,results$Def,
                            joint=TRUE,prob=0.03)
```

The returned object here is organised similarly to the single-contributor object, with the dependence on contributor removed from the organising hierarchy, as shown in Figure 9.2.

If there are three unprofiled contributors to the CSP, the function will return either a genotype list for each contributor if joint=FALSE or a genotype list with six columns (two alleles for each contributor) if joint=TRUE. If there is only a single contributor to the CSP, the results will be identical regardless of the value of

```
[[1]]
[[1]][[1]]
[[1]][[1]]$D2
[[1]][[1]]$D2$genotypes
[1] "17" "20"

[[1]][[1]]$D2$probabilities
[1] 0.9460538


[[1]][[1]]$D21
[[1]][[1]]$D21$genotypes
[1] "29"   "32.2"

[[1]][[1]]$D21$probabilities
[1] 0.8806254


[[1]][[1]]$TH01
[[1]][[1]]$TH01$genotypes
        [,1] [,2]
[1,] "6"    "9.3"
[2,] "8"    "9.3"
[3,] "6"    "8"

[[1]][[1]]$TH01$probabilities
[1] 0.3113742 0.3109442 0.2923899


[[1]][[2]]
[[1]][[2]]$D2
[[1]][[2]]$D2$genotypes
        [,1] [,2]
[1,] "17" "24"
[2,] "20" "24"

[[1]][[2]]$D2$probabilities
[1] 0.3745314 0.2656215


[[1]][[2]]$D21
[[1]][[2]]$D21$genotypes
        [,1] [,2]
[1,] "29" "31"
```

```
[2,] "31" "32.2"

[[1]][[2]]$D21$probabilities
[1] 0.5306968 0.2493529


[[1]][[2]]$TH01
[[1]][[2]]$TH01$genotypes
        [,1] [,2]
[1,] "6"   "9.3"
[2,] "8"   "9.3"
[3,] "6"   "8"
[4,] "9.3" "9.3"

[[1]][[2]]$TH01$probabilities
[1] 0.2776487 0.2232152 0.2017299 0.1318768



$topGenotypes
$topGenotypes$genotypes
$topGenotypes$genotypes[[1]]
        [,1] [,2]
D2    "17" "20"
D21   "29" "32.2"
TH01  "6"  "9.3"

$topGenotypes$genotypes[[2]]
        [,1] [,2]
D2    "17" "24"
D21   "29" "31"
TH01  "6"  "9.3"


$topGenotypes$probabilities
$topGenotypes$probabilities[[1]]
[1] 0.2594117

$topGenotypes$probabilities[[2]]
[1] 0.05518619
```

Figure 9.1   An example output of get.likely.genotypes *for a two-contributor, three-locus CSP, returning marginal genotype probabilities. The first section of the results shows the single-locus genotypes for single contributors and is broken into two subsections, one for each contributor subject to drop-out (designated with* [[1]][[1]] *for the first contributor and* [[1]][[2]] *for the second contributor). Each of these subsections is then further divided into locus sections (designated* $locusName*), which are then each split into genotypes and probabilities for that locus and that contributor. The probabilities correspond to the rows of the genotypes matrices. The second section (designated* $topGenotypes*) shows the most probable whole-profile genotype for each contributor and is split into genotypes and probabilities subsections, which are both further subdivided into contributors subject to drop-out (e.g.* $topGenotypes$probabilities[[2]] *indicates the probability of the most probable genotype for the second contributor, which is displayed in* $topGenotypes$genotypes[[2]]*).*

```
$joint
$joint$D2
$joint$D2$genotypes
     [,1] [,2] [,3] [,4]
[1,] "17" "24" "17" "20"
[2,] "20" "24" "17" "20"
[3,] "24" "25" "17" "20"
[4,] "23" "24" "17" "20"
[5,] "19" "24" "17" "20"

$joint$D2$probabilities
[1] 0.35124052 0.23496616 0.07399688 0.07061620 0.06122807


$joint$D21
$joint$D21$genotypes
     [,1] [,2]   [,3]   [,4]
[1,] "29" "31"   "29"   "32.2"
[2,] "31" "32.2" "29"   "32.2"
[3,] "30" "31"   "29"   "32.2"
[4,] "31" "32.2" "29"   "29"

$joint$D21$probabilities
[1] 0.49921332 0.16146173 0.08830646 0.08789113


$joint$TH01
```

```
$joint$TH01$genotypes
     [,1] [,2]  [,3] [,4]
[1,] "6"  "9.3" "8"  "9.3"
[2,] "8"  "9.3" "6"  "9.3"
[3,] "9.3" "9.3" "6"  "8"
[4,] "6"  "8"   "6"  "9.3"
[5,] "6"  "9.3" "6"  "8"
[6,] "6"  "6"   "8"  "9.3"

$joint$TH01$probabilities
[1] 0.15636400 0.15062049 0.13187676 0.11295039 0.10912104 0.09935616


$topGenotypes
$topGenotypes$genotype
     [,1] [,2]  [,3] [,4]
D2   "17" "24"  "17" "20"
D21  "29" "31"  "29" "32.2"
TH01 "6"  "9.3" "8"  "9.3"

$topGenotypes$probability
[1] 0.02741748
```

Figure 9.2   An example output of get.likely.genotypes *for a two-contributor CSP, obtaining joint genotype probabilities. The first section (*$joint*) displays the joint genotype probabilities at each locus for those genotypes with probability greater than* prob *(*$joint$locusName$genotypes*), as well as their associated probabilities (*$joint$locusName$probabilities*). Once again the probabilities correspond to rows in the genotypes matrices. The second section (*$topGenotypes*) displays the most likely whole-profile joint genotype (*$topGenotypes$genotype*) as well as its associated probability (*$topGeno-types$probability*).*

joint, although they will be displayed slightly differently (if joint=FALSE, the dependence on contributor will still be displayed, but as there is only one contributor, this has no real effect).

9.2 Specifics of the package

9.2.1 The parameters

Some parameters are defined in terms of a reference individual who is X under H_d. Under H_p, the reference individual is Q if Q is subject to drop-out, otherwise the first K subject to drop-out, if any, otherwise the first U, if any. If there are no contributors subject to drop-out, then there is no need for a reference individual – the parameters defined in terms of this individual are not used.

The 'nuisance' parameters, which must be eliminated under each multi-locus likelihood before taking their ratio, are

- the drop-out rates (one per replicate) for the reference individual;

- the contributions of DNA, relative to that from the reference individual (one parameter for each contributor subject to drop-out other than the reference individual); the relative contribution from an individual is used to compute drop-out rates – see below;

- the parameters of the drop-out model: locus adjustment (one per locus), power parameter (one) and degradation parameters (one for each contributor subject to drop-out);

- a drop-in parameter (optional, see below).

likeLTD maximises a (penalised) likelihood over these parameters using the R DEoptim function.

9.2.2 Key features of likeLTD

The following are some key features of likeLTD:

- It can accept 'uncertain' allele calls, in addition to present/absent, which mitigates the 'cliff edge' effect of calls that are restricted to present/absent.

- It combines information across all DNA profiling runs, thus avoiding the need for a 'consensus' profile [Gill et al., 2000].

- It allows a different drop-out rate for each contributor in each profiling runs.

- The drop-out probability for a given dose of DNA, relative to unit dose (for the reference individual), uses the model of [Tvedebrink et al., 2009].

- Drop-out rates can increase with fragment length, based on the model of Tvedebrink et al. [2012].

- As a consequence of estimating the drop-out rate for contributors, a potential contributor can be considered in hypotheses without implying that their DNA is present, because the contribution of DNA from that individual can be estimated at zero.

- Because the penalised likelihoods are maximised over the nuisance parameters, combining information over alleles, loci, replicates and individuals, there is little need for external calibration data. This is only required for a few hyperparameters – the parameters of the penalty functions and a rate parameter of the drop-out model (for which we use results from Tvedebrink et al. [2009], see below). The underlying parameters are allowed flexibility to best fit the CSP data under each hypothesis, constrained by penalty functions that depend on these hyperparameters.

- likeLTD does not use peak height information directly; it is used indirectly by the forensic scientist when making the present/uncertain/absent designations. Peak heights would provide more information and, hence, greater statistical efficiency for many CSPs but typically require extensive calibration data specific to the profiling protocol used for the CSP. Because DNA evidence is so powerful, statistical efficiency is not usually an urgent priority; robustness is usually more important and the results below show that likeLTD has good robustness properties. In the presence of multiple replicates, there is often very little loss of information from using present/uncertain/absent rather than peak heights, but if only a single profiling run is available, the loss of information can be substantial.

9.2.3   Maximising the penalised likelihood

To compute the LR, it is necessary to deal with the 'nuisance' parameters under each hypothesis. These are the drop-out model parameters $D(1)$ (one per replicate), the α_s (one per locus) and β; the contributions of DNA, relative to the reference individual, the degradation parameters γ_i (one of each for every contributor subject to drop-out); and possibly a drop-in parameter c (see above). likeLTD seeks to maximise a penalised likelihood over these parameters, with penalties on α_s, β, γ_i and c. These penalties can be thought of as prior distributions, but we do not use a Bayesian approach since we maximise over unknown parameters rather than integrate. The primary purpose of the penalty is to avoid the maximisation algorithm from exploring unrealistic regions of the parameter space.

We use the R DEoptim function to maximise the penalised likelihoods, which is a genetic algorithm utilising differential evolution optimisation to find the global minimum of a function (since we want to maximise and not minimise, it is necessary to multiply the log-likelihoods by −1).

The results from DEoptim consist of two lists, DEoptim and member. DEoptim consists of four parts:

bestmem. The parameter values that gave the maximum likelihood.

bestval. The negative log likelihood at these parameter values.

nfeval. The number of evaluations of the objective function that were carried out during optimisation.

iter. The number of generations for the optimisation.

member consists of

lower. The lower bounds of the parameters used.

upper. The upper bounds of the parameters used.

bestvalit. A vector containing the maximum likelihood at each generation.

bestmemit. A matrix containing the parameters that gave the maximum likelihood at each generation.

pop. The set of parameter values generated for the last generation.

storepop. The sets of parameters generated for previous generations.

See the DEoptim help page for more information.

Rather than estimating the nuisance parameters, they can be set to fixed values if desired by setting the upper and lower bounds. For example,

```
tofix = "dropout"
value = 0.2
# Create index of which parameters are dropout parameters
index = grep(tofix,names(paramsP$upper))
# Set those parameters upper value to the fixed
  # value for prosecution
paramsP$upper[index] = rep(value,times=length(index))
# Set those parameters lower value to the fixed
  # value for prosecution
paramsP$lower[index] = rep(value,times=length(index))

# Repeat for the defence uppers and lowers
index = grep(tofix,names(paramsD$upper))
paramsD$upper[index] = rep(value,times=length(index))
paramsD$lower[index] = rep(value,times=length(index))
```

9.2.4 Computing time and memory requirements

The Hammer case analysis described above required 49 min to run, while a test with three unprofiled contributors under H_d required 117 min. These timings were made on a desktop computer with 15 Gb of RAM and an eight core Intel i7 processor (at 3.1 GHz/core). Computing times may vary across machines. Most desktop computers will have enough memory to run all cases except when drop-in is modelled, and there are three unknown contributors under H_d. In that case, up to 1 Gb of RAM may be required per locus, depending on the number of alleles, and so three unknowns + drop-in analyses will only be possible on large-memory machines. A future version will implement an accurate approximation that greatly reduces both runtimes and RAM requirements.

The number of generations used for optimisation by DEoptim is not fixed in advance but is determined by the function evaluate. Each step is a separate optimisation, with a geometric pattern in the crossing over rate (see DEoptim:: DEoptim.control) and convergence tolerance. Each step checks for convergence after every 75 generations; convergence occurs if $\log_{10}(|1 - L_c/L_{c-75}|) \le t$ where L denotes likelihood, c the current generation and t the tolerance for the given step.

The major determinant of runtime is the number of steps required by evaluate runs for convergence, n. This is computed as $n = \max(4, 4\lceil \log_2(\bar{x}) \rceil) + r_d$ where r_d is the number of rcont parameters handed to DEoptim under the defence case and $\bar{x}$ is the mean of σ_p and σ_d, where σ_p is the standard deviation of the likelihood for the first phase (iterations 1–75) of the first step of optimisation for the prosecution (σ_d is the same, but for defence). We can see that the minimum number of steps is 4 and that the number of steps increases with the difficulty of optimisation.

A key parameter determining the DEoptim runtime is the size of the population (number of random starts per generation). Larger values tend to generate higher likelihoods, but there is a rapidly diminishing benefit with increasing population size, while runtime increases approximately linearly. likeLTD sets the population size to be four times the number of parameters varied in the optimisation (NP). For the Hammer timings above, this implies a population size of 80, while for the three-contributor test, the population size was 72. NP, and hence population size and run time, depends on the number of contributors subject to drop-out, since there is an effective amount of DNA and a degradation parameter for each of these. Runtime is also affected by the number of unprofiled contributors: these are always subject to drop-out, and summation over their possible genotypes adds to run time and to the memory requirement.

9.3 Verification

The behaviour of the WoE[1] in relation to the inverse match probability (IMP) (see Section 8.4) was used to verify the validity of likeLTD [Steele et al., 2014a]. The WoE was shown to increase towards the IMP with additional replicates in one-, two- and three-contributor settings, while never exceeding the IMP, supporting both the validity of the mathematical model underpinning likeLTD and its correct implementation in the software. Interestingly, in all cases tested, the WoE with multiple replicates exceeded the WoE that would be obtained if we had perfect information about every contributor in the sample (termed the mixLR); this suggests that multiple low-template replicates can give more information than a single good-template profile when peak height information is not used in the analysis.

These trends can be seen in Figure 9.3, taken from Steele et al. [2014b], which shows laboratory-based (a) and simulation-based (b) results for the two-contributor analyses. Eight replicates of a two-person mixture containing DNA from Donors A and C were created for each panel. Each mixture was a major–minor mixture, where

---

[1] Here, WoE is the $\log_{10}(LR)$ with large values representing strong evidence against H_d.

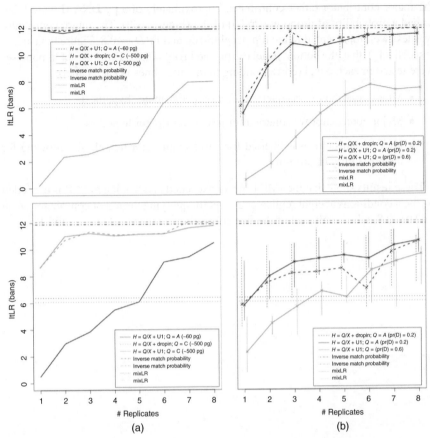

Figure 9.3   The low-template likelihood ratio (ltLR) computed by likeLTD *for two-contributor CSPs profiled at up to eight replicates. (a) Laboratory-based replicates, with the DNA template from the minor contributor greater in the lower panel (see legend boxes). (b) Simulation-based replicates, with the minor contributor having reduced drop-out in the lower panel. The simulated CSPs were generated from the profiles of Donors A and C, and the line type on the graph indicate whether the queried individual (Q) is A or C (see legend boxes). Solid lines indicate a two-contributor analysis, with the non-queried contributor regarded as unknown (U1). Dashed lines indicate a one-contributor analysis that also allows for drop-in (only for Q the major contributor). The inverse match probabilities and mixLRs are shown with horizontal lines, with font indicating the contributor. In the legend boxes, H indicates the hypotheses with X an unknown alternative to Q, and Pr(D) indicates the probability of drop-out (adapted from Figure 2 of Steele et al. [2014a]).*

the major contributor either contributed approximately 500 pg DNA (Figure 9.3a) or had a probability of drop-out ($\Pr(D)$)) of 0.2 (Figure 9.3b). The minor contributor was less well represented in the top panel (30 pg and $\Pr(D) = 0.8$, Figure 9.3a and b, respectively), than the bottom panel (60 pg and $\Pr(D) = 0.6$). Three sets of hypotheses were tested in each panel (corresponding to the three lines in each panel):

- Minor contributor was queried for a two-contributor hypothesis.

- Major contributor was queried for a two-contributor hypothesis.

- Major contributor was queried for a two-contributor hypothesis, allowing for the possibility of drop-in.

These results show that the IMP is never exceeded and that the mixLR is exceeded by most conditions at eight replicates except for querying the minor contributor when $\Pr(D) = 0.8$, where only one of the five simulations remains below the mixLR.

10

Other approaches to weight of evidence

We have seen that the likelihood ratio (LR) approach is very powerful. No matter what unusual circumstances arise in a new case: identical twins, inbreeding, fraud or drop-out, the LR provides us with a framework for assessing the evidence. For every piece of evidence, a juror should ask two questions:

- How likely is the evidence if the defendant Q is guilty?
- How likely is the evidence if another individual X is guilty?

Consequently, expert witnesses should provide as much information as possible to help answer these questions.

Although elegant and powerful, the weight-of-evidence theory based on LRs is often viewed as complicated and unfamiliar. Real crime cases are complicated, so, to some extent, it is inevitable that a satisfactory theory of evidential weight cannot be very simple. We briefly introduce alternative approaches that seem simpler but have difficulties.

10.1 Uniqueness

Match probabilities for a good-quality single-contributor crime scene profile (CSP) are often extremely small: for the 16-locus short tandem repeat (STR) system currently used in the United Kingdom, calculated match probabilities are usually less than 10^{-20}, even after a generous F_{ST} adjustment ($F_{ST} = 0.03$). When a forensic scientist reports match probabilities this small, it seems effectively equivalent to saying that he or she is reasonably certain that Q's DNA profile is unique in the population

Weight-of-Evidence for Forensic DNA Profiles, Second Edition.
David J. Balding and Christopher D. Steele.
© 2015 John Wiley & Sons, Ltd. Published 2015 by John Wiley & Sons, Ltd.
Companion Website: www.wiley.com/go/balding/weight_of_evidence

of possible sources of the CSP. If so, would not jurors be better assisted by the expert giving a 'plain English' statement of this, rather than a match probability whose unfamiliar magnitude may overwhelm or confuse? For example, perhaps an expert witness could assert that, excluding identical twins and laboratory/handling errors, in his/her opinion, the DNA profile of Q matches the CSP and there is unlikely to be another match anywhere in the world.

Although attractive in some respects, a practice of declaring uniqueness in court does lead to difficulties. One of these is how to deal with complex CSPs, involving multiple contributors, low DNA template and/or degradation, in which uniqueness cannot reasonably be asserted. Another barrier to declaring uniqueness is the problem of the non-DNA evidence in a case. The event that a particular DNA profile is unique is either true or false, no 'objective' probability can be assigned to it. Nevertheless, since this truth or falsity cannot be established in practice, a probability of uniqueness based on the information available to an expert witness, such as that obtained from DNA profile databases, together with population genetics theory, may potentially be useful to a court. The problem then arises as to what data and theory the expert should take into account. Specifically, the non-DNA evidence in a case may be directly relevant, yet it may not be appropriate for the DNA expert to assess this evidence.

Consider a crime scene DNA profile that is thought to be so rare that an expert might be prepared to assert that it is unique. Suppose that for reasons unrelated to the crime, it is subsequently noticed that the CSP matches that of the Archbishop of Canterbury. On further investigation, it is found to be a matter of public record that the Archbishop was taking tea with the Queen at the time of the offence in another part of the country. A reasonable expert would, in the light of these facts, revise downwards any previous assessment of the probability that the CSP was unique. However, this is just an extreme case of the more general phenomenon that any evidence in favour of a defendant's claim that he is not the source of the CSP is evidence against the uniqueness of his DNA profile.

10.1.1   Analysis

Suppose that the evidence E includes the fact that the DNA profile of Q matches the CSP, and let H_Q and U_Q denote the events, respectively, that Q is the source of the CSP and that the DNA profile of Q is unique in $\mathcal{P}$. It is also assumed that Q is known to have no identical twin and that laboratory or handling errors do not occur. Under these assumptions, U_Q implies H_Q, and so $P(U_Q|E) \leq P(H_Q|E)$. In fact, using (3.3), we have

$$P(U_Q|E) = P(U_Q|H_Q, E)P(H_Q|E) > \frac{P(U_Q|H_Q, E)}{1 + \sum_{X \in \mathcal{P}} w_X R_X}, \qquad (10.1)$$

where R_X denotes the LR for alternative possible culprit X, while w_X measures the weight of the non-DNA evidence against X, relative to its weight against Q.

It seems reasonable to suppose that E has no bearing on U_Q if H_Q is assumed, so that $P(U_Q|H_Q, E) = P(U_Q|H_Q)$. Further, non-uniqueness is the union of the events 'the

DNA profile of X matches the DNA profile of Q', for all $X \in \mathcal{P}$. From the elementary rules of probability, the probability of this union is bounded above by the sum of all the match probabilities (which equals the expected number of individuals in $\mathcal{P}$ sharing the DNA profile of Q). It follows that

$$P(U_Q|H_Q, E) = P(U_Q|H_Q) > 1 - \sum_{X \in \mathcal{P}} R_X. \tag{10.2}$$

Substituting (10.2) in (10.1) gives

$$P(U_Q|E) > \frac{1 - \sum_{X \in \mathcal{P}} R_X}{1 + \sum_{X \in \mathcal{P}} w_X R_X}. \tag{10.3}$$

In some cases, the non-DNA evidence does not favour Q, so that on the basis of this evidence, no individual is regarded as a more plausible suspect than Q. Then $w_X \geq 1$, for all $X \in \mathcal{P}$, and (10.3) leads to

$$P(U_Q|E) > \frac{1 - \sum_{X \in \mathcal{P}} R_X}{1 + \sum_{X \in \mathcal{P}} R_X} > 1 - 2 \sum_{X \in \mathcal{P}} R_X. \tag{10.4}$$

To help motivate the bound (10.4), consider the following simplified example. A person is sampled anonymously and at random from a population $\mathcal{P}$ of size $N + 1$ and found to have DNA profile D. The DNA profiles of the other N individuals are unknown; each is assumed to be D independently with probability p, so that the probability that D is unique in $\mathcal{P}$ is $P(U_Q) = (1 - p)^N$. A second individual is then drawn at random from $\mathcal{P}$. With probability $1/(N + 1)$, the second individual is the same as the first (call this event H_Q). Now, the second individual is typed and found to have profile D (call this observation E). If H_Q holds, E provides no additional information about U_Q, so that $P(U_Q|H_Q, E) = (1 - p)^N$. Otherwise, the two individuals sampled are distinct yet both have profile D, implying that U_Q is false. By Bayes theorem, we can calculate

$$P(H_Q|E) = \frac{1/(N + 1)}{1/(N + 1) + Np/(N + 1)} = \frac{1}{1 + Np},$$

and hence,

$$P(U_Q|E) = P(U_Q|H_Q, E)P(H_Q|E) = \frac{(1 - p)^N}{1 + Np} > 1 - 2Np,$$

which is the bound given by (10.4) in this setting.

In practice, the most important contribution to the sum in (10.3) comes from relatives of Q. See Balding [1999] for further details and discussion.

10.1.2 Discussion

Under certain assumptions, including $F_{ST} = 0.02$ (see Section 5.2) and reasonable bounds on the numbers of close relatives, simulations indicate that 11 STR loci almost always suffice to achieve a 99.9% probability of uniqueness [Balding, 1999].

However, the assumptions include that $w_X \leq 1$ for every alternative source X, which effectively implies that there is no evidence in favour of Q. There is sometimes evidence favouring the defendant, and it is not appropriate for the forensic scientist to pre-empt the jurors' assessment of the non-scientific evidence.

Focussing on the directly relevant issue, whether or not Q is the source of the CSP, rather than uniqueness, makes more efficient use of the evidence and, properly presented and explained to the court, can suffice as a basis for satisfactory prosecutions. An estimate of $P(U_Q|E)$ may also provide useful information for courts, provided that a satisfactory way is found to explain the underlying assumptions.

10.2   Inclusion/exclusion probabilities

The concept of a 'random man' has at least two meanings:

- informally, it means something like 'nobody in particular' or 'it could be anyone';

- in scientific usage, it means chosen according to a randomising device, such as a die, or a computer random-number generator.

Use of the idea of 'random man' in the first sense is generally harmless, though may cause some confusion. Serious errors can arise when 'random man' is used in the second sense in assessing weight of evidence. It is important to keep in mind that in any crime investigation, 'random man' is pure fiction: nobody was actually chosen at random in any population, and so probabilities calculated under an assumption of randomly sampled suspects have no direct bearing on evidential weight in actual cases.

The concept of 'random man' is sometimes used to calculate a CPI (combined probability of inclusion), sometimes reported as its complement the *exclusion probability*) and also referred to as the RMNE (random man not excluded) probability. When identification is at issue, a court may be informed that a sequence of tests has been conducted on samples from both crime scene and defendant and that Q was not excluded by these tests from being the source of the CSP. Instead of LRs, the expert witness advises the court on the CPI.

Note that the concept of inclusion probability can be used to summarise the average performance of a test, which not only ignores the profile of Q (as does the CPI) but is also not specific to any particular CSP. Given a choice between profiling systems, all other factors being equal, we would prefer to implement the system with the lowest average inclusion probability (equivalently, highest exclusion probability). However, this probability has little role in assessing the weight of evidence in a crime investigation, because it does not take account of any information specific to the particular crime.

10.2.1   Identification: single contributor

In the case of a single contributor CSP, the CPI is the probability that X (unrelated to Q) has no allele that is not in the profile of Q. It is numerically very similar to R_X for X unrelated to Q (with $F_{ST} = 0$), but the two approaches are conceptually different.

The rationale underlying the CPI, based on a randomly chosen X, cannot adequately cope with X related to Q and also faces severe difficulties in the presence of drop-out (Chapter 8). The idea of a random alternative suspect can lead jurors to ignore the role of the *number* of possible culprits in evidential assessments, and clear thinking about typing errors and the effect of searches can also be undermined. All of these aspects are readily dealt with in the LR framework.

One specific difficulty that 'random man' can cause in this setting concerns the argument over which population the man is supposed to have been randomly drawn from (see, e.g. Roeder et al. 1994). Since random man is a fiction, these arguments can never be resolved. The issue is important, since too broad a definition of the population leads to overstatement of the evidence, because a large population must contain many people sharing little ancestry with Q. If we try to avoid this overstatement by specifying the narrowest possible population, we are led to the population consisting of Q only, in which the match probability is 1.

10.2.2   Identification: multiple contributors

Even more important differences between CPI and LRs arise for CSPs with multiple unknown contributors (Section 6.5). Here, the CPI is the probability that an unrelated X has both alleles at each locus included among the alleles of the mixed CSP. Thus the CPI does not take account of the profile of Q, other than noting that it falls within a (usually large) class of profiles. Its advantages include being relatively easy to calculate and explain, and that it does not require any assumption about the number of contributors to the mixture.

However, ignoring relevant information can have adverse consequences. Because of the loss of information, the CPI is usually larger than the LR, often considerably so. This statistical inefficiency is sometimes seen as a virtue, in that use of CPI is said to be 'conservative' (see Section 6.3). However, conservativeness is not guaranteed (see Exercise 10.3 below), and by using an appropriate F_{ST}-value, we can make the LR conservative while still relatively efficient. Another serious disadvantage of CPI arises in a case involving two co-defendants each accused of being a contributor to the mixed CSP. The evidence can weigh more heavily against one defendant than the other, whereas the CPI will be the same for both (again, see Exercise 10.3 below).

10.2.3   Paternity

Another clear distinction between the CPI and LR approaches arises in paternity testing (Section 7.1). Consider a scenario in which a child's paternal allele at a locus, say allele C, has population proportion p_C. The CPI is the probability that X has a copy of the child's paternal allele. Making some simplifying assumptions, this probability is $1 - (1 - p_C)^2 = 2p_C - p_C^2$, irrespective of $\mathcal{G}_Q$.

In contrast, the LR has already been given at (7.4):

$$R_X = \begin{cases} p_C & \text{if } Q \text{ is homozygous for C,} \\ 2p_C & \text{if } Q \text{ has just one copy of C.} \end{cases}$$

The CPI lies between p_C and $2p_C$. If $p_C > 0.5$, then $R_X < 1$ for Q heterozygous, and so this observation *reduces* the probability that Q is the father (see Section 7.1.3). This is reasonable because an allele drawn from Q is less likely to match the child's paternal allele than is an allele drawn at random in the population. Thus, this evidence points away from Q as the father and towards any unprofiled, unrelated, man X.

The CPI approach wrongly counts the evidence in the above scenario against Q, just the same as if he had genotype CC. Although $p_C > 0.5$ is rarely realistic for STR profiling, this extreme scenario highlights the logical problems associated with not answering the relevant question. For further discussion, see Kaye [1989].

10.2.4 Discussion

The CPI seems to be attractive as a measure of evidential weight in place of LRs. There is, however, no theory linking it with the question of whether Q is a source of the CSP. This, in itself, may not be troubling if they satisfied some informal notion of fairness. This is the case in many settings, but we have seen that the CPI can sometimes lead astray. One key weakness is that all evidence decreases the CPI and, hence, counts *against* the defendant. Inclusion probabilities may have some uses in measuring and conveying evidential weight, but they should not be used without checking that they do not conflict with the logical analysis based on LRs combined using Bayes theorem.

10.3   Hypothesis testing †

In science, weight of evidence is often assessed via significance levels and/or p-values. The jury in a criminal case must reason from the evidence presented to it, to a decision between the hypotheses:

$$G \ : \ Q \text{ is guilty;}$$

$$I \ : \ Q \text{ is not guilty.}$$

Within the hypothesis-testing framework, the legal maxim 'innocent until proven guilty' would imply that I should be the null hypothesis, and so the probability of a match under I can be interpreted as a p-value.

But how do we calculate a p-value taking into account the possibility that the culprit could be a relative of the defendant? The usual answer is to invoke 'random man' and assume that hypothesis I implies that Q has been chosen randomly in some population of innocent suspects. But since no random sampling really took place, it is impossible to specify the population, as we noted above.

The hypothesis testing framework faces further difficulties with complications that we have seen are readily handled using the weight-of-evidence formula (3.3):

- How can the p-value, assessing the DNA evidence, be incorporated with the other evidence? What if the defendant produces an apparently watertight alibi? What if more incriminating evidence is found?

- How should the possibility of laboratory or handling error be taken into account?

- What if the defendant was identified only after a number of other possible culprits were investigated and found not to match?

Perhaps the most important weakness is the first: the problem of incorporating the DNA evidence with the other evidence. Hypothesis tests are designed to make accept/reject decisions on the basis of the scientific evidence only, irrespective of the other evidence. Legal decision makers must synthesise all the evidence, much of it unscientific and difficult to quantify, in order to arrive at a verdict.

The rationale behind classical hypothesis testing is based on imagining a long sequence of similar 'experiments'. Roughly speaking, a p-value addresses the question: 'how often would I observe data like this in many similar experiments if the null hypothesis were true?'. For example, in Section 5.4.1, we discuss hypothesis tests for deviations from Hardy–Weinberg equilibrium (HWE). Here, the null hypothesis of HWE is simple: it specifies precise genotype proportions as functions of allele fractions. The tests, one way or another, answer the question: if we had drawn samples from many populations in HWE and observed the same allele counts as the observed sample each time, how many times would the genotype counts deviate from the HWE expectations as much as, or more than, the observed sample?

The courtroom setting for DNA identification is very different. The null hypothesis of the defendant's innocence is complex – there are usually many alternative scenarios to that of the prosecution, and different match probabilities may be appropriate. Much confusion has been caused by trying to shoehorn the decision problem faced by jurors into the scientific hypothesis-testing framework, see Balding and Donnelly [1995] for a discussion. The second report on DNA evidence of the US National Research Council adopted this framework for most of its statistical deliberations, and consequently, some of its conclusions are flawed, see Section 11.5.

10.4   Other exercises

Solutions start on page 194.

10.1   In the original island problem of Section 2.2.1, we calculated a probability of 50% for the defendant's guilt. Assuming no evidence other than the Υ evidence, what is the probability that Q is the only Υ-bearer on the island?

10.2   Assume the following population proportions for the four main blood groups:

Blood group:	O	A	B	AB
Population proportion:	46%	42%	9%	3%

Assume that the blood groups of distinct individuals are independent. When considering the identification of the source of a blood sample drawn from a single individual, what are

(a) the inclusion probability of the typing system based on these blood groups?

(b) the case-specific inclusion probabilities for each of the four possible blood groups?

Briefly explain the uses of the probabilities calculated in (i) and (ii).

10.3 Consider a single-locus genetic typing system at which a mixed DNA sample recovered from a crime scene has profile ABD, while a man Q suspected of being a contributor to the sample has genotype BD. The population allele proportions are $p_A = 0.2$, $p_B = 0.1$ and $p_D = 0.15$, and genotype proportions equal Hardy–Weinberg values (Section 5.4). We assume here that none of the actual or possible contributors to the CSP is related and that $F_{ST} = 0$.

(a) What is the value of the LR comparing the hypothesis that Q and an unknown man were the (only) contributors to the mixture, with the hypothesis that two unknown men were the sources of the CSP?

(b) What is the inclusion probability?

(c) Repeat (a) and (b) with $G_Q = BB$.

(d) Outline (without any calculations) how your answers in (a) and (b) would be altered if it were known that the mixture had three contributors?

10.4 † Suppose that Q has been charged with an offence due to being the only one of 1000 individuals included in a database whose DNA profile matched the CSP. Consider the hypothesis:

> H_0: The profiles recorded in the database were drawn from individuals chosen at random in the total population of 10 000 mutually unrelated possible culprits.

Assume that each innocent individual has probability 10^{-6} of having a matching profile and that the profiles of different individuals are independent. The DNA profiles of the 9000 individuals not in the database are unknown.

(a) What is the significance level of the test that rejects H_0 if at least one match is observed of the CSP with a database profile?

(b) What is the relevance of your answer to (a) for the strength of the case against Q?

11

Some issues for the courtroom

Our hope is that forensic scientists reading this book will, if they have persevered this far, have gained useful insights about how to quantify the weight of DNA evidence and that these insights will help with many aspects of preparing statements and presenting them in court. If any such reader was hoping that we would be able to prescribe a formulaic approach to reporting DNA evidence in court, which satisfies the needs of jurors and the demands of judges in every case, then he/she will be disappointed. We have no magic formula to overcome the difficult issues that arise in presenting complex scientific evidence to non-expert judges and juries. However, we believe that a sufficiently deep understanding of the principles can help an expert witness to make well-informed judgements about what a clear-headed juror needs to know in order to perform his or her task. It must be up to the individual forensic scientist, reflecting their context, to find good solutions to the problem of satisfying the (partly contradictory) goals of, for example, clarity, precision, fairness, exhaustiveness and simplicity.

11.1 The role of the expert witness

As we noted in Section 3.3.1, any individual can assess the probability $P(H_Q|E_d, E_o)$ that the defendant is a source of the CSP, based on the DNA evidence E_d and any background information E_o that they feel appropriate. A juror's reasoning is, however, constrained by legal rules. For example, although it may be reasonable to believe that the fact that a person is on trial makes it more likely that they are guilty, a juror is prohibited from reasoning in this way (to avoid double counting of evidence). It is, therefore, usually regarded as inappropriate for an expert witness to report to the court their own assessment of the probability, say that Q is the source CSP (see Section 6.1).

Weight-of-Evidence for Forensic DNA Profiles, Second Edition.
David J. Balding and Christopher D. Steele.
© 2015 John Wiley & Sons, Ltd. Published 2015 by John Wiley & Sons, Ltd.
Companion Website: www.wiley.com/go/balding/weight_of_evidence

The primary role of expert witnesses is to advise the court on appropriate values for the R_X for various X, leaving the jurors to weigh these values together with the other evidence. Although jurors are not required to reason within the logical framework provided by probability theory, it may be helpful for an expert witness to give some explanation of this framework so that the option is fully available to them. The endorsement of any particular values for the w_X should usually be avoided, since this involves the juror's assessment of the non-DNA evidence and is thus usually outside the domain of the expert witness.

Although there are in principle many different likelihood ratios (LRs), in practice, it may suffice to report only a few important values. These might include the values corresponding to a brother of the defendant, another close relative such as a cousin, a person apparently unrelated to the defendant but with a very similar ethnic background and a person completely unrelated to the defendant (see Section 6.2).

11.2   Bayesian reasoning in court

The principles that we have set out here are based on the use of Bayes' theorem to assess evidence. Arguments over the acceptability of explicit Bayesian reasoning in court have been the subject of academic debate at least since the critique of Tribe [1970] provoked by Finkelstein and Fairley [1970a] (see also the authors' response, Finkelstein and Fairley, 1970b). The prospect of formalised reasoning with numbers seems not to have been welcomed by the courts. Indeed, we will see below that the England and Wales Court of Appeal have made several judgments apparently rejecting Bayesian reasoning in court. While it is regrettable that rational reasoning is officially prohibited from English courtrooms, it is also clear that most of the statistical and population genetics analyses described in this book are not appropriate for direct presentation to juries.

The critics of Bayesian reasoning in court have some support. There is a substantial academic literature pointing to the conclusion that lay jurors are prone to confusion when presented with evidence in the form of numbers. In particular, forms of words attempting to encapsulate LRs have been highlighted for their potential to sow confusion in jurors' minds; see, for example, Thompson [1989] and Koehler [1993, 1996, 2001].

These criticisms should be treated as encouragements to be cautious. On the other hand, now that DNA evidence is routinely quantified numerically, and there may often be substantial additional evidence of a non-numeric form; there seems to be no better proposal on offer to deal with the complexities that this raises than to start with a logical analysis via the weight-of-evidence formula (3.3), or similar version of Bayes' theorem. Many scientists – we will see below (see Section 11.5) even members of the US National Research Council panel – have become muddled about what scientific issues are relevant to a juror's decision, and logical reasoning has a central role to play in clarifying these issues, even if such an analysis is not directly presented in court. Our own attempts to convey evidential weight in court have been based on this formula, but we rarely utter the words 'likelihood ratio' or 'Bayes' theorem' in court.

In the interests of jurors unfamiliar with formalised reasoning, we believe that the intuition given by the mathematical formalism can, and should, be conveyed in ways that avoid its explicit use. We see the theory developed in this book as crucial for good understanding by the forensic scientist and not as a prescription for how to convey this understanding in court.

Assuming here that R_X for unrelated X was reported to be 1 in a million, one form of words that we have used in court to help convey the weight of evidence is as follows:

> 'Consider the hypothetical scenario in which, on the basis of the non-DNA evidence, a juror considered that there were 1000 men who could be the questioned DNA source: Mr Q and 999 men unrelated to him. If each is initially considered equally likely to be the source, the effect of the DNA profiling results would be to change the probability that Mr Q is indeed the correct source from 1 in 1000 (or 0.1%) up to 99.9%'

This does not address the issue of X related to Q.

Another common strategy is to develop a hierarchy of expressions to try to convey the weight of different ranges of LR values. It may be helpful to the court to express weight of evidence on a logarithmic scale, similar to the Richter scale for earthquakes. The following verbal descriptions of the support given by an LR to one proposition relative to another have been proposed [UK Association of Forensic Science Providers, 2009]:

Verbal description	Limited	Moderate	Moderately strong	Strong	Very strong	Extremely strong
$-\log_{10}(\text{LR})$	0–1	1–2	2–3	3–4	4–6	> 6

Unlike the Richter scale, it is important to note that the LR is not an absolute measure of weight of evidence but depends on the hypotheses chosen. For criticisms of this approach, see Martire et al. [2013].

11.3   Some fallacies

11.3.1   The prosecutor's fallacy

The 'prosecutor's fallacy' [Thompson and Schumann, 1987] is a logical error that can arise when reasoning about DNA profile evidence. You may be aware of the error in elementary logic of confusing 'A implies B' with 'B implies A'. For example, if A denotes 'is a cow' and B denotes 'has 4 legs', then (ignoring rare anomalies) A implies B, but the converse does not hold. The prosecutor's fallacy is similar to this logical error but is in terms of probabilities.

The fallacy consists of confusing $P(A \mid B)$ with $P(B \mid A)$.[1] If it is accepted that the probability that the criminal is very tall is 90%, it does not follow that the next very tall man you meet has 90% probability of being the criminal. The correct way of obtaining $P(A|B)$ from $P(B|A)$ is given by the appropriate version of Bayes' theorem, for example, the weight-of-evidence formula (3.3).

Transcripts of actual court cases have in the past very often recorded statements that indicate that the match probability is being confused with the probability that the defendant is innocent. For example,

- 'I can estimate the chances of this semen having come from a man other than the provider of the blood sample … less than 1 in 27 million';

- 'The FBI concluded that there was a 1 in 2600 probability that the semen … came from somebody other than Martinez'.

In these quotations, an expert witness has made a statement about the probability that the defendant is not the source of the crime profile. Such statements lie outside the domain of an expert witness, are logically distinct from a match probability or LR and appear to be instances of the prosecutor's fallacy. A correct version of these statements should involve a conditional statement of the form '*if* the defendant were not the source, then the probability … '. The fallacy can be extremely detrimental to defendants and has led to successful appeals in the United Kingdom (e.g. Section 11.4.1).

In addition to clear instances of the fallacy, there are many ambiguous phrases that could suggest the fallacy to jurors. For example, a sentence of the form

'The probability that an innocent man would have this profile is 1 in 1 million'

is ambiguous because it is unclear whether this refers to a match probability for a particular man (probably OK) or to the probability that there exists (anywhere) an innocent, matching man (probably fallacious).

For further examples and discussion of the fallacy, see Koehler [1993], Balding and Donnelly [1994], Evett [1995] and Robertson and Vignaux [1995]. Good [1995] exposed an instance of a similar error of reasoning with probabilities. In the US trial of O.J. Simpson, a member of the defence team pointed out that few wife-batterers go on to murder their wives, drawing the conclusion that Simpson was a priori unlikely to be guilty. Good pointed out that this ignores the fact that Simpson's ex-wife had been murdered and the available data suggest that a high proportion of murdered women who have been battered by their husbands were in fact murdered by the husband.

11.3.2   The defendant's fallacy

Another error of logic that can arise in connection with DNA evidence usually favours the defendant and is consequently dubbed the 'defendant's fallacy'. Suppose that a

---

[1] The prosecutor's fallacy is a particular case of the 'error of the transposed conditional'.

crime occurs in a nation of 100 million people and a profile frequency is reported as 1 in 1 million. The fallacy consists of arguing that since the expected number of people in the nation with a matching profile is 100, the probability that the defendant is guilty is at most only 1 in 100, or 1%.

This conclusion would be valid only if, ignoring the DNA evidence, every person in the nation is equally likely to be the culprit. In the notation of the weight-of-evidence formula, each w_χ is equal to 1. In practice, such an assumption is rarely reasonable. Even if there is little directly incriminating evidence beyond the DNA profile match, there is always background information presented in evidence, such as the location and nature of the alleged offence, that will make some individuals more plausible suspects than others.

11.3.3 The uniqueness fallacy

Consider a country of population n, and a DNA profile with match probability for an unrelated, alternative culprit calculated to be less than $1/n$. Ignoring relatives, coancestry and any other evidence, this implies a mathematical expectation of less than 1 for the number of matching individuals in the population. The fallacy is to conclude that an expectation less than 1 means that no such individual exists. In the UK case *R. v. Gary Adams* (Section 11.4.3), a match probability of 1 in 27 million had been reported and the judge concluded:

> ' ... I should think that there are not more than 27 million males in the United Kingdom, which means that it is unique.'

When the expectation of a count is very small, it is approximately equal to the probability that the count is not zero. Thus, the match probability would have had to have been several orders of magnitude smaller than 1 in 27 million in order for the judge's conclusion to be reasonable.

11.4   Some UK appeal cases

11.4.1   Deen (1993)

Deen[2] was the first conviction based primarily on DNA evidence to come before a UK appeal court. The judgement upholding the appeal is notable for establishing the prosecutor's fallacy as an important issue in the presentation of DNA evidence.

At the trial, an expert witness agreed with the statement:

> 'the likelihood of [the source of the semen] being any other man but Andrew Deen is 1 in 3 million',

a clear example of the fallacy. There were further ambiguous statements which could have been interpreted as instances of the fallacy. The appeal judgement explained

---

[2] *The Times*, January 10, 1994.

the fallacy at length and stated explicitly the judges' view that it was not necessary for them to assess how much difference the misrepresentation of the DNA evidence would have made.

11.4.2   Adams (1996)

In April 1996, the Court of Appeal in London overturned the conviction for rape of Denis Adams.[3] The case was unusual in that, although the prosecution case was supported by a DNA profile match linking the defendant with the crime, for which a match probability of 1 in 200 million was reported; the defence case also had strong support, from an alibi witness who seems not to have been discredited at trial, but most importantly from the victim, who stated in court that the defendant did not look like the man who attacked her, nor did he fit the description that she gave at the time of the offence.

Thus, jurors were faced with the task of weighing the strongly incriminating DNA evidence against substantial non-scientific evidence pointing in the other direction. Experts for prosecution and defence agreed that the logically satisfactory way to do this was via Bayes' theorem (in effect, the weight-of-evidence formula (3.3)). The defence presented numerical illustrations of the theorem applied sequentially to each of the principal pieces of evidence. The jury was reminded repeatedly that they should assess their own values. In his summing up, the trial judge reminded jurors that they were not obliged to use Bayes' theorem.

An appeal was launched for reasons other than the principle of using Bayes' theorem, but although not raised by the appellant, in upholding the appeal, the court expressed strong reservations as to whether such evidence should have been allowed to go before the jury. They added that as they had not heard argument on the matter, their views should be interpreted as provisional.

It is difficult for our non-legal minds to find much of substance in the appeal court judgement. The judges express concern that in advising jurors of the accepted, rational method of reasoning with probabilities the Bayes' theorem evidence:

> ' ... trespasses on an area peculiarly and exclusively within the province of the jury, namely the way in which they evaluate the relationship between one piece of evidence and another.'

The substantive paragraph seems to the one which says

> ' ... the [Bayes' theorem] methodology requires ... that items of evidence be assessed separately according to their bearing on the accused's guilt, before being combined in the overall formula. That in our view is far too rigid an approach to evidence of the nature which a jury characteristically has to assess, where the cogency of ... identification evidence

---

[3] [1996] 2 Cr.App.R. 467. See also *The Times*, May 9, 1996; *New Scientist*, June 8, 1996 and December 13, 1997.

> may have to be assessed … in the light of the strength of the chain of evidence of which it forms part.'

This is somewhat unclear, but it seems that the judges have misunderstood the sequential nature of Bayes' theorem, correctly applied (Section 3.3.1). They go on to say

> ' … the attempt to determine guilt or innocence on the basis of a mathematical formula, applied to each separate piece of evidence, is simply inappropriate to the jury's task. Jurors evaluate evidence and reach a conclusion not by means of a formula, mathematical or otherwise, but by the joint application of their individual common sense and knowledge of the world to the evidence before them.'

The issue of how common sense and knowledge of the world equip jurors to understand the weight of DNA evidence, presented as a match probability, was not elaborated by the judges.

The successful appeal led to a retrial at which Bayes' theorem was introduced again (the higher court's opinion was not binding and the retrial judge chose to ignore it). In fact, the court went further and, with the collaboration of prosecution and defence, distributed to the jury a questionnaire that guided them through the application of Bayes' theorem, together with a calculator. Adams was convicted again and appealed again. The Court of Appeal, in dismissing the second appeal,[4] took the opportunity to reinforce its earlier opinion:

> ' … expert evidence should not be admitted to induce juries to attach mathematical values to probabilities arising from non-scientific evidence … '

Clearly, jurors may be confused by unfamiliar mathematical formalism, and indeed, it seems that the judge did make errors in his summing up of the Bayes' theorem evidence. Nevertheless, once DNA evidence is presented in terms of probabilities, it seems hard to sustain the argument that jurors should be *prevented* from hearing an explanation of the accepted method of reasoning with probabilities to incorporate it with the non-DNA evidence. The appeal court judgements give no hint as to how jurors might otherwise be guided in this task.

11.4.3   Doheny/Adams (1996)

These two related appeals[5] were against convictions based to varying extents on DNA evidence. The court upheld one appeal and dismissed the other; the prosecutor's fallacy was an issue in both cases, but there were additional issues concerning the calculated match probabilities.

---

[4] [1998] 1 Cr.App.R 377.

[5] [1997] 1 Cr App R 369. See also *The Times*, August 14, 1996. Note that this 'Adams' is different from the one discussed above.

The appeals are important because the Court of Appeal took the opportunity that they offered to give directions as to how DNA evidence should be presented at trial:

> '[the scientist] will properly explain to the Jury the nature of the match.... He will properly, on the basis of empirical statistical data, give the Jury the random occurrence ratio – the frequency with which the matching DNA characteristics are likely to be found in the population at large. Provided that he has the necessary data, and the statistical expertise, it may be appropriate for him then to say how many people with the matching characteristics are likely to be found in the United Kingdom – or perhaps in a more limited relevant sub-group, such as, for instance, the Caucasian sexually active males in the Manchester area.
>
> This will often be the limit of the evidence which he can properly and usefully give. It will then be for the Jury to decide, having regard to all the relevant evidence, whether they are sure that it was the Defendant who left the crime stain, or whether it is possible that it was left by someone else with the same matching DNA characteristics.'

The term 'random occurrence ratio' introduced by the court appears to be a synonym for match probability. This novel coinage is an unwelcome addition to the many terms already available.

The court then suggested a model for summing up DNA evidence, which implies assessing the DNA evidence before taking any other evidence into account:

> 'Members of the jury, if you accept the scientific evidence called by the Crown, this indicates that there are probably only four or five white males in the UK from whom that semen stain could have come. The Defendant is one of them. The decision you have to reach, on all the evidence, is whether you are sure that it was the Defendant who left that stain or whether it is possible that it was one of the other small group of men who share the same DNA characteristics.'

This approach is not illogical, but it encounters several difficulties. The court does not seem to have considered the effect of relatives on evidence, nor the problem of integrating the DNA evidence with the other evidence. However, an expert can take relatives into account when calculating the expected number of matches and can emphasise that the calculation does not take the non-DNA evidence into account.

With current match probabilities, the expected number of matching individuals in the United Kingdom is usually a tiny fraction, which may be difficult for jurors to interpret. It may be preferable to use probabilities; for example,

> 'There is only one chance in a hundred that any other person in the UK shares the crime-scene profile.'

Although not without merit, the court's prescription for the presentation of DNA evidence has had unsatisfactory consequences in its attempts to limit what a forensic scientist may comment on. In addition to the phrase 'That will often be the limit … ' quoted above, the court went on to say

> The scientist should not be asked his opinion on the likelihood that it was the defendant who left the crime stain, nor when giving evidence should he use terminology which may lead the jury to believe that he is expressing such an opinion.'

It seems clear from the context that the Court of Appeal was trying to restrain expert witnesses from any danger of falling into the prosecutor's fallacy, but in some UK courts, these words seem to have been used to straightjacket forensic scientists. For a further discussion on its implications, see Lambert and Evett [1998].

11.4.4   Watters (2000)

The appellant had been convicted of burglary on the basis of short tandem repeat (STR) profiles obtained from cigarette ends obtained at the crime scenes, together with reports of similarities between these crimes.

A forensic expert reported matches at six STR loci. The match probability for an unrelated individual was said to be 1 in 86 million, whereas that for a brother was 1 in 267. It was reported in evidence that the defendant had two brothers and that one of these had been arrested in connection with the offences but released without charge and without a DNA sample having been taken from him (or the other, unarrested, brother). The forensic scientist agreed that her results did not prove that the defendant was the source of the crime-scene DNA and also that DNA evidence should not be used in isolation and without other supporting evidence. The supporting evidence provided by the prosecution included that the defendant was a smoker and lived in the general locality of the burglaries.

The defence requested the trial judge to, in effect, rule that there was no case to answer because the brothers had not been excluded. He rejected this argument and invited the jury to consider the 1 in 267 match probability for brothers. His arguments for rejecting the defence request include the existence of the supporting evidence about locality, and because the defence had not supplied the names, addresses and dates of birth of the brothers. The latter argument would seem to contravene the principle that a defendant does not have to prove anything.

The appeal court found that ruling to be wrong, on the basis of a mixture of valid and misguided reasoning. The court drew attention to the forensic scientist's admission that a DNA profile match does not constitute proof of identity and contrasted DNA evidence with fingerprint evidence where certain identification is routinely reported in court. This contrast is spurious, and it is regrettable that the attempt to assess probabilities for error in the case of DNA profile evidence was interpreted by a senior court as implying that DNA evidence is inferior to fingerprint evidence: comparable difficulties for fingerprint evidence have routinely been swept under the carpet (see Section 4.9).

The court was on stronger ground in asserting that although the DNA evidence was very strong in excluding a source of the DNA unrelated to the defendant, the existence of unexcluded brothers, one of whom had been suspected of the offences, cast doubt on whether the correct brother was in court. In our view, the court should be congratulated for taking the issue of the unexcluded brothers seriously, which courts in many earlier trials failed to do. However, its interpretation of 'certainty' is doubtful: two brothers each with match probability less than 1 in 200 could, in the absence of other evidence and other possible culprits, correspond to a posterior probability of guilt of over 99%, which a reasonable juror may accept as 'certainty' within the meaning of the law.

Whatever the merits of that argument, it seems absurd to argue, as the Court of Appeal did in its judgement, that further STR testing which resulted in the defendant continuing to match, but now with a match probability for a brother of 1 in 29 000, was still insufficient for a conviction because this much smaller match probability still does not suffice to eliminate the possible brother. On this basis, they did not order a retrial. The implication of the court's argument seems to be that any uncertainty quantified numerically fails to achieve 'proof'. Uncertainty prevails for all forms of evidence; the attempt to quantify should be seen as a sign of the strength of DNA evidence, not a weakness.

11.4.5   T (2010)

The disputed evidence consisted of shoe marks found at the crime scene, and a pair of shoes in the possession of Mr T that may have been the source of those marks. To assess this possibility, the forensic scientist computed an LR comparing:

G : The marks came from the shoes recovered from Mr T.

I : The marks came from a different pair of shoes.

The computation used a database of 8122 shoes and included subjective allowances for wear and damage. Specifically, there was a feature of the marks that may have been caused by a stone lodged in the shoe, whereas no such stone was found in the recovered shoes, and these showed greater wear than was evident from the marks. No numerical LR was reported at trial, instead the expert witness reported 'a moderate degree of scientific support for [hypothesis G] ...' based on a standard verbal scale (page 167).

The England and Wales Court of Appeal (full transcript: EWCA Crim 2439, 2010) ruled that an LR was not appropriate in this setting and instead sought to limit the expert witness to stating that 'this mark could have come from this shoe'. Its ruling could be used to challenge future attempts to try to put forensic evidence on a scientific footing by using evaluations based on data and quantitative reasoning rather than subjective opinion based on experience. DNA evidence remains a special case: rational, scientific assessment of evidence is permitted in English courts for DNA, if not for other forms of scientific evidence.

Although aspects of the evidence presented in this case can be criticised and were subject to good and bad criticisms by the Court, it made no attempt to scrutinise the obvious deficiencies of its preferred approach. A statement that a mark could have come from a shoe is of little value, because of its reliance on the state of mind of the examiner: some examiners will interpret 'could have' in limited terms, justified only when there are substantial observations to support it, whereas other experts will interpret the same phrase more liberally: from a sceptical viewpoint, any mark could have come from any shoe unless there are strong grounds to rule this out. Moreover, the approach does not rely on a systematic method that can be assessed and improved; there may be little opportunity for a defence to understand the reasoning behind the expert's subjective judgment and, hence, to propose legitimate criticisms.

The Court put great faith in the ability of experts to evaluate evidence based only on experience, failing to recognise that without training and assessments based on blind proficiency trials, experience alone is of little value: 20 years of performing a task do not imply that the task was always done well. Moreover, it does not help in understanding the principles of evidence evaluation because there is little opportunity in casework to know the truth and, hence, critically assess performance. Similar points are made by Rudin and Inman [2010], who noted the dangers inherent in memory biases that can affect the aspects of an expert's experience that are presented to a court and the fact that experience-based evidence, being personal, is essential unrebuttable.

Although the Court correctly highlighted problems with the forensic evidence in this case, it went beyond criticising those problems and tried to rule out progress towards a rational, scientific approach to evidence evaluation. Such an approach would rely on explicit data and assumptions that give the defence appropriate opportunities to probe weaknesses. One weak argument put forward by the Court was that a database of 8122 shoes was not large enough relative to the 42 million shoes sold in the United Kingdom each year; it also correctly noted that usage of shoes varies over regions and years, which may not be reflected in the database. The proposed database, which is imperfect like almost all such databases, appears to provide the best available data relevant to the question at hand. However, the Court chose to make no use of the database, appearing to imply that no data is preferable to limited data.

The prosecutor's fallacy (Section 11.3.1) was made by the Court: when referring to the prosecution hypothesis the ruling talked about 'the probability that the Nike trainers owned by the appellant had made the marks'. Attentive readers of this book will recognise that the relevant issue is 'the probability of the marks if the Nike trainers owned by the appellant had made them'.

For further discussion of the ruling and the issues raised, see Aitken et al. [2011], Berger et al. [2011], Redmayne et al. [2011], Robertson et al. [2011], Morrison [2012] and Sjerps and Berger [2012].

11.4.6   Dlugosz (2013)

This England and Wales Court of Appeal judgment (EWCA Crim 2., 2013) related to three cases, the first of which was R v. Dlugosz. In each case, mixed low-template DNA (LTDNA) profiles were obtained that appeared to support the prosecution allegation, but the DNA expert witness for the prosecution was unable to present what the court referred to as a 'random match probability'; since no match arose, this term is not appropriate, and we understand it to mean an LR.

Therefore, in each case, complex evidence was presented without any numerical evaluation of evidential weight. The DNA evidence was instead accompanied by subjective, verbal opinions of evidential weight that do not lie on a scale of increasing strength. For example, in one case, the expert witness only sought to say that the defendant could have been a contributor to the crime-scene DNA profile and accepted that this statement implied that he might not have been a contributor. In another case, the expert said that based on her experience, it was 'rare' to observe so many matching alleles if the suspected contributor was not in fact a true contributor. However, the 'rarity' of such an observation depends strongly on the number of observed alleles. Moreover, each complex electropherogram (epg) has individual characteristics that can make general statements unreliable. For example, in this case, peaks attributable to the alleged contributor showed a high degree of height variation.

When challenged, the experts justified their conclusions on the basis of substantial casework experience and some proficiency testing but not necessarily in conditions closely matching the circumstances of the case at hand. As in R v. T (Section 11.4.5), the Court of Appeal placed great store in the value of experience, but in reality, as we noted above, experience alone is of little value. Evett and Pope [2013] expressed the point well, saying

> The ruling assumes that a scientist who has quantitatively evaluated a large number of mixtures cases will have the knowledge to assign a reliable qualitative opinion of weight of evidence in a case that is too complex for a quantitative assessment. However, there is no scientific basis for this belief – no scientific literature provides a reliable methodology, scientists are not trained to make such assessments and there is no body of standards to support them. Casework experience is not a substitute. The true composition of the DNA result in any given case cannot be known so it does not provide a reliable control for learning purposes.

They went on to propose 'a national programme of assessing scientists' interpretation of specially constructed DNA mixtures under controlled conditions against agreed standards'. However, a problem remains that complex DNA profiles can be complex in so many different ways that an expert may misjudge some complex profiles even though they have shown proficiency in evaluating others.

At the time of the original trials, it may have been reasonable to argue that statistical methods and software to evaluate complex LTDNA profiles were not sufficiently well established or widely available for the courts to rely on, but by the time of the Court of Appeal ruling that was no longer the case (see Chapter 8). The Court paid almost no attention to the real solution to its problem: validated software that can provide objective, quantitative assessments of even very complex profiles, taking into account many sources of variation such as degradation, DNA contribution from different contributors and replicate-specific effects.

See [Gill, 2014b, p. 122] for further discussion and also Champod [2013] for a discussion of DNA transfer that mentions some issues raised in this ruling and in other cases.

11.5   US National Research Council reports

The report on DNA evidence of the US National Research Council [1992] was intended to settle some of the controversy surrounding DNA evidence. The report devoted almost no space to weight-of-evidence issues and espoused no principles for assessing evidence. In response to the perceived problem of population hetero-geneity, it proposed an *ad hoc* solution, the 'ceiling principle', which involves using the largest frequency for each DNA profile band, assessed from a diverse range of populations. The report suffered a hostile reception from many sides. In particular, the ceiling principle was criticised, many taking the view that it was too generous to defendants, and others complaining that it was a poorly considered response to a poorly specified problem.

Following pressure from, principally, the Federal Bureau of Investigation (FBI), a new committee was formed and a new report was issued [National Research Council, 1996]. This time there was a (very brief) discussion of evidential weight. Unfortu-nately, the report adopted the principle that weight of evidence is measured in terms of an arbitrary and unjustified assumption that the defendant was sampled randomly in some population (see the discussion on 'random man' in Section 10.2). Consequently, the second report is also flawed in many important respects. Perhaps the most aston-ishing feature is the serious error on the issue of a defendant identified by a database search, mentioned already in Section 3.4.5. The report recommends that the weight of DNA evidence in this setting be measured by the profile frequency multiplied by the size of the database, indicating much weaker evidence than in a no-search case.

The rationale for this recommendation is clearly flawed. The report compares the DNA search scenario to tossing 20 coins repeatedly, so that, eventually, it becomes likely that the outcome 'all heads' will have occurred. This analogy is misleading: in the DNA search case, we know that somebody, somewhere, has the profile. Moreover, as we noted in Section 2.3.4, the relevant question is not 'what is the probability of observing a match?' but 'given that I have observed a match, how strong is the evi-dence against this individual?'. The correct analysis of the database search scenario was discussed in Section 3.4.5; the number of profiles in the database is essentially irrelevant to evidential weight, and the NRC's recommendation is similar to suggest-ing that a good answer be multiplied by an arbitrary number.

Aside from the database search blunder, which gives a large and unwarranted benefit to some defendants, the report otherwise errs consistently in favour of prosecutions. Although the committee which prepared the report was chaired by one eminent population geneticist and included several others, failure to adopt appropriate weight-of-evidence principles means that the population genetics theory was misapplied. Because it considered only profile proportions for 'random man', the report fails to address the key population genetics issue of the correlations between the DNA profile of the defendant and those of other possible culprits. Instead, the report gives undue attention to the less important issue of within-person genetic correlations, and recommendations substantially favouring prosecutions are the result.

Recommendation 4.3, concerning small and isolated groups such as Native American tribes for which little relevant data is available, is particularly troubling and could lead to miscarriages of justice. Similarly troubling is the report's recommendation that relatives be considered only when *there is evidence* that they are possible culprits. We saw in Section 3.4.3 how this approach can be misleading. Moreover, this assumption in effect reverses the burden of proof: it should be for the prosecution to prove that the defendant committed the offence and, hence, that his relatives did not, rather than for the defence to show that relatives are plausible suspects. Similarly unfair to defendants is the recommendation that subpopulation issues be taken into account only if *all* the alternative possible culprits are from the same subpopulation as the defendant. See Balding [1997] for further discussion and criticism of the report.

11.6   Prosecutor's fallacy exercises

Smith has a genetic type that matches that of blood found at a crime scene. The match probability for an individual unrelated to Smith is reported to be 1 in 1000. The statements below are adapted from Evett [1995]. Discuss whether each is probably OK, probably an instance of the prosecutor's fallacy, or ambiguous.

11.1   The probability that the profile would be of this type if it had come from someone other than Smith is 1 in 1000.

11.2   The chance that a man other than Smith left blood of this type is 1 in 1000.

11.3   The probability that someone other than Smith would have blood of this type is 1 in 1000.

11.4   The evidence is 1000 times more likely if Smith were the source of the blood than if an unrelated man were the source.

11.5   It is 1000 times more probable that Smith is the source of the blood than that an unrelated man is the source.

11.6   The chance of a man other than Smith having the same blood type is 1 in 1000.

Solutions to exercises

Section 2.4, page 19

2.1 (a) Manchester:

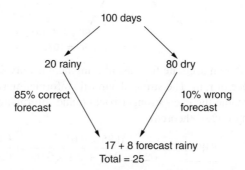

Out of every 25 rainy days, on average 17 had been correctly forecast as rainy, whereas 8 had been wrongly forecast to be dry. The probability of rain, given a forecast of rain, is equal to the proportion of rainy days on which rain had been forecast, which is $17/25 = 68\%$.

In Chapter 3, we used mathematical formalism and Bayes' theorem (3.8) to compute this answer more concisely. Here is a preview of a more formal derivation. Let

$$W \equiv \text{'rainy'}$$

$$\overline{W} \equiv \text{'dry'}$$

$$F \equiv \text{'rain forecast'}.$$

We seek $P(W \mid F)$. Now

$$P(W) = 0.2, \qquad P(F \mid W) = 0.85,$$
$$P(\overline{W}) = 0.8, \qquad P(F \mid \overline{W}) = 0.1,$$

Weight-of-Evidence for Forensic DNA Profiles, Second Edition.
David J. Balding and Christopher D. Steele.
© 2015 John Wiley & Sons, Ltd. Published 2015 by John Wiley & Sons, Ltd.
Companion Website: www.wiley.com/go/balding/weight_of_evidence

and by Bayes' theorem, we have

$$P(W \mid F) = \frac{P(F \mid W)P(W)}{P(F \mid W)P(W) + P(F \mid \overline{W})P(\overline{W})}$$

$$= \frac{0.85 \times 0.2}{0.85 \times 0.2 + 0.1 \times 0.8} = \frac{0.17}{0.25} = 0.68.$$

(b) Alice Springs:

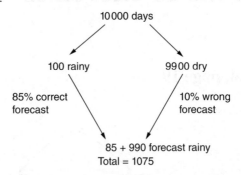

So after an accurate forecast of rain, there is only 85/1075, just under 8%, chance of rain. The strong diagnostic information (forecast of rain) cannot overcome the even stronger background information that rain is unlikely. Using Bayes' theorem,

$$P(W \mid F) = \frac{0.85 \times 0.01}{0.85 \times 0.01 + 0.1 \times 0.99} = \frac{0.0085}{0.1075} \approx 0.079.$$

2.2 (a) Because of the extremely strong background information that the condition is very rare, after an accurate diagnosis of the condition, there is still only 992/20 990, just under 5%, chance of having it. We derive this first using Bayes' theorem:

$$P(W \mid F) = \frac{0.992 \times 0.0001}{0.992 \times 0.0001 + 0.002 \times 0.9999}$$

$$= \frac{0.0000992}{0.0000992 + 0.0019998} = \frac{992}{20\,990} \approx 0.047,$$

and now represent the derivation using a diagram:

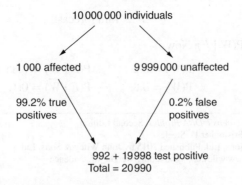

(b) Using Bayes' theorem,

$$P(W \mid F) = \frac{0.992 \times 0.004}{0.992 \times 0.004 + 0.002 \times 0.996}$$

$$= \frac{0.003968}{0.003968 + 0.001992} = \frac{3968}{5960} \approx 0.67.$$

Thus, although the background evidence implies that the person tested is unlikely to have the condition (probability $1/250 = 0.004$), the diagnostic information from the accurate test result is so strong that it overcomes this and results in a 2/3 probability that the person does have the condition.

2.3 (a) Use of the island problem formula (2.1) gives

$$P(G \mid E) = \frac{1}{1 + N \times p} = \frac{1}{1 + 1000 \times 5 \times 10^{-6}} = \frac{1}{1.005} \approx 0.995.$$

Thus, although there are 1000 alternative possible culprits on the island, because they are all unrelated to Q, and the match probability in this case is very small, under the assumptions of the island problem, the observation that Q has Υ reduces the total probability of guilt for all alternatives from $1000/1001$ to $1/201$. Conversely, the probability that Q is the culprit has increased from $1/1001$ to $200/201$ or about 99.5%.

(b) Since the islanders are initially equally under suspicion but have different levels of relatedness to Q and, hence, different match probabilities, formula (3.7) applies to calculate the probability of guilt:

$$P(G \mid E) = \frac{1}{1 + \sum_{i=1}^{N} m_X}$$

$$= \frac{1}{1 + 0.015 + 20 \times 2 \times 10^{-4} + 979 \times 5 \times 10^{-6}}$$

$$= \frac{1}{1 + 0.015 + 0.004 + 0.004895}$$

$$= \frac{1}{1.023895} \approx 0.977.$$

Under these assumptions, despite the Υ match, there remains probability almost 2.5% that Q is not the culprit, about five times higher than when we assumed that there are no relatives of Q on the island.

(c) Under the assumptions of the island problem, the observation that b does not have Υ excludes him from suspicion, and hence, the corresponding term vanishes from the match probability formula, leaving

$$P(G \mid E) = \frac{1}{1 + 0.004 + 0.004895} = \frac{1}{1.08895} \approx 0.991.$$

(d) The juror should use whatever background information is available to assess plausible ranges for the numbers of relatives in each category, and the probability of different numbers in each range. This might entail using everyday knowledge to assess typical numbers of siblings, nephews/uncles, cousins, etc. In principle, the weight-of-evidence formula should then be used with each plausible set of numbers of relatives, and the resulting values averaged according to the probabilities of each set. In practice, at the cost of an error in favour of Q, it is often satisfactory to consider only an upper bound on the plausible number of relatives in each category, for example, 10 siblings and 100 cousins.

Section 3.6, page 38

3.1 (a) The juror needs to assess the probability that Q would have glass fragments with refractive index matching the crime-scene fragments if in fact Q is innocent and X is the culprit and also the same probability if Q did commit the crime.

(b) Different individuals X might have different probabilities of carrying glass fragments of a specific type on their clothing according to their trade, hobbies and lifestyle. For individuals X who live in the same dwelling as Q, or other close associates of Q, the relevant probabilities may be affected by the fact that Q has such fragments and the degree of contact between X and Q. This is because of the possibility of cross-contamination by body contact or exchange of clothing.

(c) The following are some possible answers:

   (i) Surveys of the frequency with which glass fragments are found on the clothing of individuals in the general population.

   (ii) Surveys of the relative frequencies of different refractive indices on clothing.

   (iii) Evidence about the frequency with which glass fragments are detected on the clothing of individuals who have recently broken a window, after a time delay corresponding to that between the crime and the apprehension of the defendant and the seizure of his/her clothing.

   (iv) Estimates of the error distribution for measurements of glass fragments obtained from clothing.

3.2 (a) Including brothers:

$$P(G \mid E) = \frac{1}{1 + 2/267 + 10^5/87 \text{ million}}$$

$$= \frac{1}{1 + 2/267 + 1/870} \approx 0.991.$$

(b) Excluding brothers:

$$P(G \mid E) = \frac{1}{1 + 1/870} \approx 0.999.$$

Thus, under our assumptions, the overall case against Q is very strong even if the brothers are included and is overwhelming if they are excluded.

3.3 Let us write

E_d for the DNA evidence

E_o for the background information

(i.e. location of crime, local adult male population)

E_a for the alibi evidence

E_v for the victim's evidence

(a) If we assume that $w_X = 1$ for all 1.5×10^5 men in $\mathcal{P}$, taken here to be the population of men between 15 and 60 within 15 km of the crime scene, and $w_X = 0$ for all other individuals, and that the likelihood ratios (LRs) for E_d are $R_X = 5 \times 10^{-9}$ ($= 1$ in 200 million) for all X, then by using the weight-of-evidence formula (3.3), the probability of the defendant's guilt would be

$$P(G \mid E_d, E_o) = \frac{1}{1 + \sum_{i \in P} R_X} = \frac{1}{1 + N \times R_X}$$

$$= \frac{1}{1 + 1.5 \times 10^5 \times 5 \times 10^{-9}}$$

$$= \frac{1}{1 + 7.5 \times 10^{-4}} \approx 0.999.$$

(b) Using (3.5) and (3.6), the joint LR for two items of evidence is the product of the two LRs, but the second ratio should take into account the first piece of evidence. Here, it seems natural to assume that the alibi evidence and DNA evidence are independent, and so the overall LR is the product of the LR for the alibi evidence (4) and that for the DNA evidence (5×10^{-9}). Thus we have

$$P(G \mid E_a, E_d, E_o) = \frac{1}{1 + 1.5 \times 10^5 \times 4 \times 5 \times 10^{-9}}$$

$$= \frac{1}{1 + 3 \times 10^{-3}} = \frac{1000}{1003}$$

$$\approx 0.997.$$

(c) Reasoning the same way as in (b) and assuming that the victim's evidence is independent of both alibi and DNA evidence, we have

$$P(G \mid E_v, E_a, E_d, E_o) = \frac{1}{1 + 1.5 \times 10^5 \times 10 \times 4 \times 5 \times 10^{-9}}$$

$$= \frac{1}{1 + 3/100} = \frac{100}{103}$$

$$\approx 0.97.$$

(d)

$$P(G \mid E_v, E_a, E_d, E_o) = \frac{1}{1 + 1.5 \times 10^5 \times 10 \times 4 \times 5 \times 10^{-8}}$$

$$= \frac{1}{1 + 3/10} = \frac{10}{13}$$

$$\approx 0.77.$$

(e) There are many unreasonable assumptions, among them that Q has no relatives in the population P and that nobody outside P is a possible culprit. None of the probability assignments can be regarded as precise, but in particular there seems little basis for assigning $R_X = 4$ for the alibi evidence, rather than, say, 3 or 5, because there is little available data on the reliability of alibi witnesses and an appropriate value will be specific to the particulars of the case. (Nevertheless, jurors can consider ranges of plausible values, and making such subjective judgements on the veracity of witnesses is what jurors are there to do.) The calculations in (a) through (d) should thus be regarded as providing an aid to clear thinking about the problem rather than a definitive answer to it.

(f) The question is open ended and gives scope for individual answers. In both trials of Adams (Section 11.4.2), the defence guided jurors through a calculation similar to those made above, allowing jurors to fill in probability assignments. In both appeals, this use of mathematical formalism was severely criticised by the judges, but they could offer no reasonable alternative for conveying the relevant insights about combining different items of evidence.

3.4 (a) Formula (2.5) applies (see Section 3.4.5). The database search eliminates from suspicion the $k = 20$ individuals found not to have Y, and so

$$P(G \mid E) = \frac{1}{1 + (N - k)p} = \frac{1}{1 + 80 \times 0.01} = \frac{5}{9} \approx 0.56.$$

The effect of the search is thus to increase the probability that Q is the culprit from 50% to about 56%.

(b) Your a priori assessment can be formalised as $w_X = 1$ for every X whose Y-state is recorded in the database and $w_X = 0.1$ for all other islanders. Of the former group, all but Q are eliminated from suspicion, and so other than Q only the 80 individuals with $w_X = 0.1$ contribute to the denominator of the weight-of-evidence formula (3.3):

$$P(G \mid E) = \frac{1}{1 + \sum_{X=1}^{80} w_X R_X} = \frac{1}{1 + 80 \times 0.1 \times 0.01} = \frac{1}{1.08} \approx 0.93.$$

3.5 (a) When most of the individuals in the database are not plausible alternative suspects for a particular offence, their elimination due to non-matching with the crime-scene profile (CSP) has no effect on the case against Q. However, because some members of the database may have been considered as (perhaps remotely) possible culprits and because observation of the non-matches confirms the belief that the profile is rare, the database search always has at least a small effect of strengthening the case against the one individual who does match.

(b) The effect on the probability analysis of information such as that Q was ill and far from the crime scene is manifested through large values of w_X for at least some X. Remember that w_X is defined (page 23) as the weight of the other (i.e. non-DNA) evidence against X relative to its weight against Q. Since the non-DNA evidence here makes it very unlikely a priori that Q committed the crime, there are typically many men for whom there is no such exculpatory evidence and so for whom a very large value of w_X is appropriate. For illustration, suppose that the possible suspects include Q, thought very unlikely to have committed the crime, and say 10 000 men with $w_X = 1000$. If a DNA profile match is observed with Q, for which $R_X = 4 \times 10^{-8}$ is accepted, while the DNA profiles of the other men are not available to the court, then we would calculate

$$P(G \mid E_d, E_o) = \frac{1}{1 + 10\,000 \times 1000 \times 4 \times 10^{-8}} = \frac{1}{1.4} \approx 0.71.$$

Under these assumptions, the DNA profile evidence is so strong that it still leads to Q being more likely than not the culprit, but the effect of the large w_X is that the probability of G fails to be high enough for a satisfactory prosecution. If we had $w_X = 1$ for all 10 000 alternative suspects, instead of $w_X = 1000$, then the probability of guilt would be approximately 0.9996.

(c) A match with an individual who could not have committed the crime suggests that a close relative of this individual is the culprit, either an identical twin if one exists or else a first-degree relative.

Section 5.8, page 78

5.1 (a) By (5.3), the variance of the subpopulation allele proportions is

$$\text{Var}[\tilde{p}] = p(1 - p)F_{ST} = 0.15 \times 0.85 \times 0.02 = 0.00255,$$

and so the standard deviation (SD) is $\sqrt{0.00255} \approx 0.05$.

(b) Using the 'plus or minus two SDs' rule of thumb, the 95% interval for a subpopulation proportion is 0.15 ± 0.1, which is the interval $(0.05, 0.25)$. Using the beta distribution with $p = 0.15$ and $\lambda = 49$, the central 95% interval is $(0.066, 0.26)$. (In R, use the command qbeta(c(0.025, 0.975),0.15*49,0.85*49)).

5.2 (a) Substituting these values into (5.4) gives

$$\widehat{F_{ST}} = \frac{(11-10)^2 + (15-10)^2 + (8-10)^2 + (9-10)^2 + (12-10)^2}{5 \times 10 \times 90}$$

$$= 0.0078.$$

(b) If p is unknown, a natural estimate is the mean of the observed values: $(0.11 + 0.15 + 0.08 + 0.09 + 0.12)/5 = 0.11$. Substituting into (5.4) and replacing k with $k - 1$, we obtain

$$\widehat{F_{ST}} = \frac{(11-11)^2 + (15-11)^2 + (8-11)^2 + (9-11)^2 + (12-11)^2}{4 \times 11 \times 89}$$

$$= 0.0077.$$

5.3 Using (5.6),

(a)

$$P(3,0) = \frac{(1 - F_{ST})p}{1 - F_{ST}} \times \frac{F_{ST} + (1 - F_{ST})p}{1} \times \frac{2F_{ST} + (1 - F_{ST})p}{1 + F_{ST}}$$

$$= \frac{p(F_{ST} + (1 - F_{ST})p)(2F_{ST} + (1 - F_{ST})p)}{1 + F_{ST}}$$

(b)

$$P(2,1) = 3\frac{p(1 - p)(1 - F_{ST})(F_{ST} + (1 - F_{ST})p)}{1 + F_{ST}}.$$

When $p = 0.25$, these formulas evaluate to

F_{ST}:	0	0.02	0.1	1
$P(3,0) \times 1000$	16	19	31	250
$P(2,1) \times 1000$	141	146	165	0

5.4 The allele proportion estimates are

$$\widehat{p_B} = \frac{2 \times 5 + 15 + 10}{130} = \frac{7}{26}$$

$$\widehat{p_G} = \frac{2 \times 10 + 15 + 20}{130} = \frac{11}{26}$$

$$\widehat{p_R} = \frac{2 \times 5 + 10 + 20}{130} = \frac{8}{26},$$

and so the calculations for Pearson's goodness-of-fit statistic are

Genotype	BB	AA	CC	AB	BC	AC	Total
Observed (O)	5	10	5	15	10	20	65
Expected (E)	4.71	11.63	6.15	14.81	10.77	16.92	65
$(O-E)^2/E$	0.018	0.230	0.216	0.002	0.055	0.559	1.08

The statistic has three degrees of freedom, and since the observed value ($= 1.08$) is less than 3, we see immediately without resort to tables that this value is not significant, and we cannot reject the hypothesis of Hardy–Weinberg equilibrium (HWE).

5.5 Writing a, b, c and d for the sample counts of AB, Ab, aB and ab haplotypes respectively, Pearson's goodness-of-fit statistic for testing the hypothesis of linkage equilibrium is (page 75)

$$nr^2 = n \times \frac{(ad - bc)^2}{(a + b)(c + d)(a + c)(b + d)}$$

$$= 43 \times \frac{(80 - 100)^2}{30 \times 13 \times 15 \times 28}$$

$$= 0.11.$$

Again, the observed value of the statistic ($= 0.11$) is less than the degrees of freedom ($= 1$) and so we cannot reject the hypothesis of linkage equilibrium.

Section 6.6, page 106

6.1 (a) Consider a man U from the village, unrelated to Q. At locus 1,

$$R_u^1 = \frac{2(F_{ST} + (1 - F_{ST})p_A)(F_{ST} + (1 - F_{ST})p_B)}{(1 + F_{ST})(1 + 2F_{ST})}$$

$$= \frac{2(0.03 + 0.97 \times 0.06)(0.03 + 0.97 \times 0.14)}{(1 + 0.03)(1 + 0.06)} \approx 0.0268,$$

and at locus 2,

$$R_u^2 = \frac{(2F_{ST} + (1 - F_{ST})p_C)(3F_{ST} + (1 - F_{ST})p_C)}{(1 + F_{ST})(1 + 2F_{ST})}$$

$$= \frac{2(0.06 + 0.97 \times 0.04)(0.09 + 0.97 \times 0.04)}{(1 + 0.03)(1 + 0.06)} \approx 0.0117.$$

Using the product rule, the two-locus LR for each of the unrelated men in the village is

$$R_U = R_U^1 \times R_U^2 \approx 0.00031.$$

Now consider H, a half-brother of Q. In the notation of Section 6.2.4, for a half-brother, we have $\kappa_0 = \kappa_1 = 0.5$, and

$$M_1^1 = \frac{F_{ST} + (1 - F_{ST})(p_A + p_B)/2}{1 + F_{ST}}$$

$$= \frac{0.03 + 0.97 \times 0.1}{1.03} \approx 0.123,$$

$$M_1^2 = \frac{2F_{ST} + (1 - F_{ST})p_C}{1 + F_{ST}}$$

$$= \frac{0.06 + 0.97 \times 0.04}{1.03} \approx 0.0959.$$

By (6.6), the single-locus match probabilities for a half-brother are

$$R_H^1 = \kappa_2 + M_1^1 \kappa_1 + R_U^1 \kappa_0$$

$$\approx 0 + 0.123 \times 0.5 + 0.0268 \times 0.5 = 0.075,$$

$$R_H^2 \approx 0.0959 \times 0.5 + 0.0117 \times 0.5 = 0.054,$$

and the two-locus match probability is

$$R_H = R_H^1 \times R_H^2 \approx 0.0040.$$

Uncles have the same relationship to Q as does his half-brother, and hence, R_h also applies to the two uncles. Finally, using the weight-of-evidence formula (3.3) and noting that we have assumed $w_X = 1$ for all the alternative possible culprits, we have

$$P(G \mid E_d, E_o) = \frac{1}{1 + \sum w_X R_X} = \frac{1}{1 + 3R_H + 41R_U}$$

$$\approx \frac{1}{1 + 3 \times 0.0040 + 41 \times 0.00031} = 0.976.$$

(b) Because the migrant men are from a different population, a lower value of F_{ST} is appropriate, say $F_{ST} = 1\%$. Also, the allele proportions most appropriate for these men are those of the population from which they have come (and not, e.g. those of the suspect Q). Then,

$$R_M^1 = \frac{2(F_{ST} + (1 - F_{ST})p_A)(F_{ST} + (1 - F_{ST})p_B)}{(1 + F_{ST})(1 + 2F_{ST})}$$

$$= \frac{2(0.01 + 0.99 \times 0.08)(0.01 + 0.99 \times 0.09)}{(1 + 0.01)(1 + 0.02)}$$

$$\approx 0.0172,$$

$$R_M^2 = \frac{(2F_{ST} + (1 - F_{ST})p_C)(3F_{ST} + (1 - F_{ST})p_C)}{(1 + F_{ST})(1 + 2F_{ST})}$$

$$= \frac{2(0.02 + 0.99 \times 0.06)(0.03 + 0.99 \times 0.06)}{(1 + 0.01)(1 + 0.02)}$$

$$\approx 0.00689.$$

and the two-locus LR for each of the migrant men is

$$R_M = R_M^1 \times R_M^2 \approx 0.000118.$$

The probability of guilt for Q taking these 20 men into account in addition to the 44 previously considered is

$$P(G \mid E_d, E_o) = \frac{1}{1 + 3R_H + 41R_U + 20R_M} \approx 0.975.$$

Because of the remote possibility of coancestry with Q, the 20 migrant men have a lower match probability than the men in the village, and much less than the half-brother and uncles. Taking the migrant men into account makes very little difference to the strength of the overall case against Q.

6.2 Let m_X denote the match probability reported by the forensic scientist, which is the LR taking only the possibility of a chance match into account, and write R_X for the LR that can allow for either a chance match or fraud. The definition of R_X is (3.2)

$$R_X = \frac{P(E \mid C = X)}{P(E \mid C = Q)},$$

where E denotes the DNA evidence, and we have suppressed mention of the other evidence E_o

We assume that the evidence E is just as likely if fraud occurs as it is if Q is the culprit, and so fraud is not an issue if $C = Q$. If $C = X$, we need to consider separately the possibilities that fraud (event F) has and has not occurred. To do this, we use a basic theorem of probability known as the Theorem of Total Probability, or the Partition theorem (see any introductory probability textbook). In Bayesian statistics, it is sometimes called 'extending the conversation' because we introduce an extra variable into the probability space. The theorem gives

$$P(E \mid C = X) = P(E \mid F, C = X)P(F \mid C = X) + P(E \mid \overline{F}, C = X)P(\overline{F} \mid C = X),$$

where $\overline{F}$ denotes no error or fraud.

The probability of fraud could alter according to the identity of the culprit: we assume this not to be the case, so that $P(F \mid C = X) = 1\%$ and $P(\overline{F} \mid C = X) = 99\%$ for all X. Moreover, if $\overline{F}$ holds, then E is just as likely as in the scenario in which the possibility of fraud is ignored. Then

$$R_X = 0.01 + 0.99m_X,$$

which is ≈ 0.01 since $m_X = 10^{-9}$. Thus, under our assumptions, the precise value of the match probability is immaterial, beyond the fact that it is much less than the probability of fraud.

6.3 (a) Given that the mixed CSP and the victim's profile are both AB, the second contributor to the crime-scene DNA can have genotype AA, AB or BB. In evaluating the probabilities of these genotypes, we use the sampling formula (5.6) and the fact that we have observed, in Q and K, three A and one B alleles.

$$R_X = \frac{P(AB \mid \mathcal{G}_Q = AA, \mathcal{G}_K = AB, X \text{ and } K \text{ are the sources})}{P(AB \mid \mathcal{G}_Q = AA, \mathcal{G}_K = AB, Q \text{ and } K \text{ are the sources})}$$

$$= P(AA \mid AAAB) + 2P(AB \mid AAAB) + P(BB \mid AAAB)$$

$$= \frac{(3F_{ST} + (1 - F_{ST})p_A)(4F_{ST} + (1 - F_{ST})p_A)}{(1 + 3F_{ST})(1 + 4F_{ST})} +$$

$$\frac{2(3F_{ST} + (1 - F_{ST})p_A)(F_{ST} + (1 - F_{ST})p_B)}{(1 + 3F_{ST})(1 + 4F_{ST})} +$$

$$\frac{(F_{ST} + (1 - F_{ST})p_B)(2F_{ST} + (1 - F_{ST})p_B)}{(1 + 3F_{ST})(1 + 4F_{ST})}$$

$$= (p_A + p_B)^2 \qquad \text{if } F_{ST} = 0.$$

(b) The second contributor to the crime-scene DNA must have genotype AB, and we have previously observed one copy each of the A, B, C and D alleles. So

$$R_X = \frac{P(ABCD \mid \mathcal{G}_Q = AB, \mathcal{G}_v = CD, X \text{ and } K \text{ are the sources})}{P(ABCD \mid \mathcal{G}_Q = AB, \mathcal{G}_v = CD, Q \text{ and } K \text{ are the sources})}$$

$$= 2P(AB \mid ABCD) = \frac{2(F_{ST} + (1 - F_{ST})p_A)(F_{ST} + (1 - F_{ST})p_B)}{(1 + 3F_{ST})(1 + 4F_{ST})}$$

$$= 2p_A p_B \qquad \text{if } F_{ST} = 0.$$

(c) The required LR is now

$$R_X = \frac{P(ABCD \mid \mathcal{G}_Q = AB, X_1 \text{ and } X_2 \text{ are the sources})}{P(ABCD \mid \mathcal{G}_Q = AB, Q \text{ and } X_2 \text{ are the sources})}.$$

In the denominator of R_X, we must have $\mathcal{G}_{X_2} = CD$, whereas in the numerator, the genotypes of X_1 and X_2 can be AB,CD; AC,BD; or AD,BC; and each pair can occur in either order. Under our usual assumptions, all these genotype combinations have the same probability, and so we have

$$R_X = 12P(AB \mid ABCD) = \frac{12(F_{ST} + (1 - F_{ST})p_A)(F_{ST} + (1 - F_{ST})p_B)}{(1 + 3F_{ST})(1 + 4F_{ST})}$$

$$= 12p_A p_B \qquad \text{if } F_{ST} = 0.$$

6.4 (a) Because $F_{ST} = 0$, $\mathcal{G}_{Q1}$ can be ignored. The numerator of (6.15) is unaffected, but in the denominator, AB is the only genotype possible for X_2. The probability $2p_A p_B$ cancels to leave

$$R_X = \frac{P(ABC \mid \mathcal{G}_{Q2} = CC, X_1 \text{ and } X_2 \text{ are the sources})}{P(ABC \mid \mathcal{G}_{Q2} = CC, Q2 \text{ and } X_2 \text{ are the sources})}$$

$$= 6p_C(p_A + p_B + p_C).$$

(b) (i) The most likely genotype for the major contributor is AB, in which case R_X is just the AB match probability, which is $2p_A p_B$.

(ii) The minor contributor can have any genotype that includes at least one C allele and no allele other than A, B and C. Thus,

$$R_X = 2p_A p_C + 2p_B p_C + p_C^2.$$

Section 7.6, page 128

7.1 (a) At locus 1, ignoring mutation, the child's maternal allele is A, and hence, the paternal allele is B. Since Q has one C allele, the LR when $F_{ST} = 0$ is $2p_B$. (This exact case is not included in Table 7.1, but it is equivalent to those of rows 7 and 9 in that table.) The case of locus 2 is given in row 2 of Table 7.1: since c's paternal allele is A and Q has one A allele, the $F_{ST} = 0$ LR is $2p_A$. At locus 3, c's paternal allele is ambiguous; if Q is the father, he must have transmitted his A allele to c, but if X is the father, he could have transmitted either A or B. The LR is given in row 7 of Table 7.1: $2(p_A + p_B)$. Therefore,

$$\text{Overall LR} = 0.1 \times 0.2 \times 0.3 = 0.006.$$

(b) Using the formulas from Table 7.1, the overall LR is

$$\frac{0.1 + 1.9 \times 0.05}{1.05} \times \frac{0.1 + 1.9 \times 0.1}{1.05} \times \frac{0.1 + 1.9 \times 0.15}{1.05} \approx 0.0188.$$

(c) At locus 1, if Q is the father, then we have observed four alleles in M and Q: an A, a B and two C (the alleles of c do not count, as they are replicates of observed alleles). If X is the father, then c's allele B is distinct from that of Q and makes a fifth observed allele. The LR is twice the conditional probability of drawing a B, given the observed ABCC:

$$R_X = 2P(B \mid ABCC) = \frac{2F_{ST} + (1 - F_{ST})p_B}{1 + 3F_{ST}}$$

$$= \frac{0.1 + 1.9 \times 0.05}{1.15} \approx 0.170.$$

Similarly for locus 2:

$$R_X = 2P(A \mid AAAB) = \frac{2(3F_{ST} + (1 - F_{ST})p_A)}{1 + 3F_{ST}}$$

$$= \frac{0.3 + 1.9 \times 0.1}{1.15} \approx 0.426.$$

For locus 3, if Q is the father, we have observed AABC, whereas if X is the father, we have two equally likely possibilities, A or B, and so:

$$R_X = 2P(A \mid AABC) + P(B \mid AABC)$$

$$= \frac{2(3F_{ST} + (1 - F_{ST})(p_A + p_B))}{1 + 3F_{ST}}$$

$$= \frac{0.3 + 1.9 \times 0.15}{1.15} \approx 0.509.$$

The overall LR is the product of these three, about 0.0368.

7.2 (a) From Table 6.3, for a half-brother, $\bar{\kappa} = 1/4$, and substituting this into (7.11), we obtain the overall LR:

$$(0.25 + 1.5p_B) \times (0.25 + 1.5p_A) \times (0.25 + 1.5(p_A + p_B)) = 0.06175.$$

(b) The weight-of-evidence formula (3.3) applies to paternity problems in the same way as for identification. Since we are given that $w_X = 1$ for each alternative possible father, the posterior probability that Q is the father is

$$P(Q \text{ is father} \mid \mathcal{G}_C, Q, M) = \frac{1}{1 + 0.06175 + 10 \times 0.006}$$

$$= \frac{1}{1.12175} \approx 0.89.$$

We see from the calculation that the alternate possibilities that (i) the half-brother of Q is the father of c and (ii) a man unrelated to Q is the father are approximately equally likely.

7.3 The genotypes of c, Q and M are inconsistent with Q being the father unless a mutation has occurred. The LR for this locus when $F_{ST} = 0$ is

$$R_X = \frac{p_A}{\mu^f_{C \to A}} = \frac{0.05}{0.0005} = 100,$$

corresponding to strong evidence that Q is not the father. However, the overall LR is $0.006 \times 100 = 0.6$, so that under our assumptions, the four-locus profiles point (weakly) to Q being the father rather than any particular unrelated man, despite the apparent mutation.

7.4   (a) The profiles of Q and c do not match at any of the three loci, and so (7.13) simplifies in this case to

$$R = 0.25 + 0.5 \times R_X^p,$$

where R_X^p is the LR for paternity when the mother is unavailable. From Table 7.2, we obtain the following expressions:

Locus	s	c	R_X^p
1	BC	AB	$4(F_{ST} + (1 - F_{ST})p_B)/(1 + F_{ST})$
2	AB	AA	$2(2F_{ST} + (1 - F_{ST})p_A)/(1 + F_{ST})$
3	AC	AB	$4(F_{ST} + (1 - F_{ST})p_A)/(1 + F_{ST})$.

and hence we obtain when $F_{ST} = 0$,

$$R = (0.25 + 1/0.4) \times (0.25 + 1/0.4) \times (0.25 + 1/0.8) \approx 11,$$

and when $F_{ST} = 0.02$,

$$R = (0.25 + 0.5 \times 1.04/(0.08 + 0.98 \times 0.2))^2 \times$$
$$(0.25 + 0.5 \times 1.04/(0.08 + 0.98 \times 0.4)) \approx 6.16.$$

Thus the data support the sibling relationship over no relation. This is unsurprising since they share an allele at each locus.

(b) Recall from (7.1) that R_X^p compares the hypotheses that Q and c are unrelated (numerator) and that Q and c are father–child (denominator), whereas R compares the hypotheses that Q and c are siblings (numerator) and that Q and c are unrelated (denominator). Thus the required single-locus LR, comparing siblings with father–child, is

$$R \times R_X^p = 0.5 + 0.25 \times R_X^p.$$

Evaluating over the three loci, with $F_{ST} = 0$, we obtain

$$(0.5 + 0.05)^2 \times (0.5 + 0.1) \approx 0.18.$$

Thus, although the short tandem repeat (STR) profile data support the sibling relationship over no relation, the father–child relation is supported even more. Again, this is unsurprising since, under the father–child relationship, ignoring mutation, exactly one allele will be shared identical by descent (IBD) at each locus.

Section 10.4, page 163

10.1 The bound (10.4) gives

$$P(U \mid E) > 1 - 2 \sum_{i \in P} R_X = 1 - 2 \times 100 \times 0.01 = -1,$$

which is not very useful. However, there is an exact formula under the assumption of independent Υ-possession, which is

$$P(U \mid E) = \frac{(1 - p)^N}{1 + Np}$$

$$= \frac{(1 - 0.01)^{100}}{1 + 100 \times 0.01} \approx \frac{0.366}{2} = 0.183.$$

Before the observation that Q has Υ, the probability that the culprit is the unique Υ-bearer on the island is $(1 - 0.01)^{100}$ or about 37%. However, this observation, together with the possibility that Q is not the culprit, halves this probability to around 18%.

10.2 (a) The inclusion probability is

$$P(\text{Inc}) = p^2 + (1 - p)^2 = 1 - 2p + 2p^2,$$

where p denotes the probability of Υ possession. $P(\text{Inc}) = 1$ at both $p = 0$ and $p = 1$ and takes its optimal (i.e. minimal) value of 0.5 when $p = 0.5$. Thus, if there are only two possible outcomes for an identification test, it is best if these outcomes are equally likely. More generally, for a test with k possible outcomes, the inclusion probability has minimum value $1/k$ when all the outcomes have probability $1/k$.

  (b) (i) The inclusion probability of the test is

$$(0.46)^2 + (0.42)^2 + (0.09)^2 + (0.03)^2 = 0.397.$$

  (ii) The case-specific inclusion probabilities are equal to the individual population proportions.

The inclusion probability of the test (i) gives an average inclusion probability over many investigations and is useful for comparing this test with other possible tests but has no role in evaluating the evidence in a particular case. The case-specific inclusion probabilities (ii) do give a measure of evidential weight in a particular case, and here they are equivalent to match probabilities.

10.3 (a) Except for relabelling of alleles, the LR is the same as in the example introduced on page 104, in the case $F_{ST} = 0$,

$$R_X = \frac{12 p_B p_D (p_A + p_B + p_D)}{p_A + 2 p_B + 2 p_D} \approx 0.116.$$

(b) P(inclusion)=$(p_A + p_B + p_D)^2 = (0.45)^2 = 0.2025$.

(c) The exclusion probability is unaffected. For the LR, this is essentially the same as Exercise 4(a) of Section 6.6:

$$R_X = 6p_B(p_A + p_B + p_D) = 0.27.$$

(d) Calculating LRs becomes increasingly more complicated as the number of contributors to the sample increases, because, for example, we must consider all possible combinations of three genotypes consistent with the observed mixed profile. In contrast, the combined probability of inclusion (CPI) is unaffected by the number of contributors.

10.4 (a) The probability of at least one matching profile in the database is given by the binomial probability formula

$$P = 1 - (1 - p)^n.$$

Here, $n = 1000$ and $p = 10^{-6}$, and so $P \approx np = 10^{-3}$.

(b) If you said that this result was highly relevant, then either you are not paying enough attention or you have a fundamental disagreement with the authors. As discussed in Sections 3.4.5 and 6.1, we know H_0 to be false irrespective of the data, but in any case, whether or not the database was chosen randomly in the population has little relevance to the case against the specific individual, Q, observed to match.

Section 11.6, page 178

11.1   This is a conditional statement (probability of profile *if* it came from …) and so is probably OK.

11.2   Statement about the probability that Smith did not leave the blood: fallacy.

11.3   This is similar to the ambiguous statement discussed in the text, but it is very close to being fallacious, suggesting that there is only a 1 in 1000 chance that there is *anybody* other than Smith with the blood type.

11.4   Fairly standard expression of the LR: logically OK.

11.5   Statement about the probability that Smith is the source: fallacy.

11.6   Ambiguous, but in our opinion, less suggestive of fallacy than 3.

References

C Aitken, CEH Berger, JS Buckleton, C Champod, J Curran, AP Dawid, IW Evett, P Gill, J Gonzalez-Rodriguez, G Jackson, et al. Expressing evaluative opinions: a position statement. *Science & Justice*, 51 (1): 1–2, 2011.

C Aitken and F Taroni. *Statistics and the Evaluation of Evidence for Forensic Scientists*, 2nd ed. John Wiley & Sons, Ltd, 2004.

C Aitken, F Taroni, and P Garbolino. A graphical model for the evaluation of cross-transfer evidence in DNA profiles. *Theoretical Population Biology*, 63 (3): 179–190, 2003.

V Albanèse, NF Biguet, H Kiefer, E Bayard, J Mallet, and R Meloni. Quantitative effects on gene silencing by allelic variation at a tetranucleotide microsatellite. *Human Molecular Genetics*, 10 (17): 1785–1792, 2001.

HL Allen, K Estrada, G Lettre, SI Berndt, MN Weedon, F Rivadeneira, CJ Willer, AU Jackson, S Vedantam, S Raychaudhuri, et al. Hundreds of variants clustered in genomic loci and biological pathways affect human height. *Nature*, 467 (7317): 832–838, 2010.

MM Andersen, A Caliebe, A Jochens, S Willuweit, and M Krawczak. Estimating trace-suspect match probabilities for singleton Y-STR haplotypes using coalescent theory. *Forensic Science International: Genetics*, 7: 264–71, 2013.

KL Ayres. A two-locus forensic match probability for subdivided populations. *Genetica*, 108 (2): 137–143, 2000a.

KL Ayres. Relatedness testing in subdivided populations. *Forensic Science International*, 114 (2): 107–115, 2000b.

KL Ayres. Paternal exclusion in the presence of substructure. *Forensic Science International*, 129 (2): 142–144, 2002.

KL Ayres and DJ Balding. Measuring departures from Hardy–Weinberg: a Markov chain Monte Carlo method for estimating the inbreeding coefficient. *Heredity*, 80 (6): 769–777, 1998.

KL Ayres and DJ Balding. Paternity index calculations when some individuals share common ancestry. *Forensic Science International*, 151 (1): 101–103, 2005.

KL Ayres, J Chaseling, and DJ Balding. Implications for DNA identification arising from an analysis of Australian forensic databases. *Forensic Science International*, 129 (2): 90–98, 2002.

Weight-of-Evidence for Forensic DNA Profiles, Second Edition.
David J. Balding and Christopher D. Steele.
© 2015 John Wiley & Sons, Ltd. Published 2015 by John Wiley & Sons, Ltd.
Companion Website: www.wiley.com/go/balding/weight_of_evidence

KL Ayres and WM Powley. Calculating the exclusion probability and paternity index for X-chromosomal loci in the presence of substructure. *Forensic Science International*, 149 (2): 201–203, 2005.

DJ Balding. Errors and misunderstandings in the second NRC report. *Jurimetrics*, 37 (4): 469–476, 1997.

DJ Balding. When can a DNA profile be regarded as unique? *Science & Justice*, 39: 257–260, 1999.

DJ Balding. Interpreting DNA evidence: can probability theory help? In *Statistical Science in the Courtroom*, pages 51–70. Springer, 2000.

DJ Balding. The DNA database search controversy. *Biometrics*, 58 (1): 241–244, 2002.

DJ Balding. Likelihood-based inference for genetic correlation coefficients. *Theoretical Population Biology*, 63 (3): 221–230, 2003.

DJ Balding. Evaluation of mixed-source, low-template DNA profiles in forensic science. *Proceedings of the National academy of Sciences of the United States of America*, 110: 12241–12246 2013.

DJ Balding and J Buckleton. Interpreting low template DNA profiles. *Forensic Science International: Genetics*, 4: 1–10, 2009.

DJ Balding and P Donnelly. The prosecutors fallacy and DNA evidence. *Criminal Law Review*, 711–721, 1994.

DJ Balding and P Donnelly. Inference in forensic identification. *Journal of the Royal Statistical Society: Series A (Statistics in Society)*, 158 (1): 21–53, 1995a.

DJ Balding and P Donnelly. Evaluating DNA profile evidence when the suspect is identified through a database search. *Journal of Forensic Sciences*, 41 (4): 603–607, 1996.

DJ Balding and RA Nichols. DNA profile match probability calculation: how to allow for population stratification, relatedness, database selection and single bands. *Forensic Science International*, 64 (2): 125–140, 1994.

DJ Balding and RA Nichols. A method for quantifying differentiation between populations at multi-allelic loci and its implications for investigating identity and paternity. In *Human Identification: The Use of DNA Markers*, pages 3–12. Springer, 1995.

DJ Balding and RA Nichols. Significant genetic correlations among Caucasians at forensic DNA loci. *Heredity*, 78 (6), 1997.

DJ Balding, M Greenhalgh, and RA Nichols. Population genetics of STR loci in Caucasians. *International Journal of Legal Medicine*, 108 (6): 300–305, 1996.

DJ Balding, M Krawczak, JS Buckleton, and JM Curran. Decision-making in familial database searching: KI alone or not alone? *Forensic Science International: Genetics*, 7 (1): 52–54, 2013. ISSN: 1878-0326 (Electronic); 1872-4973 (Linking).

O Bangsø and P-H Wuillemin. Object oriented Bayesian networks a framework for topdown specification of large Bayesian networks and repetitive structures. Aalborg Universitetsforlag: Aalborg, Denmark, 2000.

M Bauer. Rna in forensic science. *Forensic Science International: Genetics*, 1 (1): 69–74, 2007.

M Bauer, I Gramlich, S Polzin, and D Patzelt. Quantification of MRNA degradation as possible indicator of postmortem interval – a pilot study. *Legal Medicine*, 5 (4): 220–227, 2003a.

M Bauer, S Polzin, and D Patzelt. Quantification of RNA degradation by semi-quantitative duplex and competitive RT-PCR: a possible indicator of the age of bloodstains? *Forensic Science International*, 138 (1): 94–103, 2003b.

S Beleza, NA Johnson, SI Candille, DM Absher, MA Coram, et al.. Genetic architecture of skin and eye color in an African-European admixed population. *PLoS Genetics*, 9 (3): e1003372, 2013.

C Benschop, C van der Beek, H Meiland, A van Gorp, A Westen, and T Sijen. Low template STR typing: effect of replicate number and consensus method on genotyping reliability and DNA database search results. *Forensic Science International: Genetics*, 5 (4): 316–28, 2011. ISSN: 1878-0326.

CEH Berger, J Buckleton, C Champod, IW Evett, and G Jackson. Evidence evaluation: a response to the court of appeal judgment in *R v T*. *Science & Justice*, 51 (2): 43–49, 2011.

JM Bernardo and AFM Smith. *Bayesian Theory*, volume 405. John Wiley & Sons, Inc., 2009.

FR Bieber, CH Brenner, and D Lazer. Finding criminals through DNA of their relatives. *Science*, 5778: 1315, 2006.

A Biedermann and F Taroni. Bayesian networks for evaluating forensic DNA profiling evidence: a review and guide to literature. *Forensic Science International: Genetics*, 6 (2): 147–157, 2012.

A Biedermann, J Vuille, and F Taroni, editors. DNA, statistics and the law: a cross-disciplinary approach to forensic inference. Volume 5 of *Frontiers in Genetics*. Frontiers Media SA, 2014.

M Bill, P Gill, J Curran, T Clayton, R Pinchin, M Healy, and J Buckleton. Pendulum – a guideline-based approach to the interpretation of STR mixtures. *Forensic Science International*, 148 (2): 181–189, 2005.

CW Birky Jr. The inheritance of genes in mitochondria and chloroplasts: laws, mechanisms, and models. *Annual Review of Genetics*, 35 (1): 125–148, 2001.

Ø Bleka, G Dørum, H Haned, and P Gill. Database extraction strategies for low-template evidence. *Forensic Science International: Genetics*, 9: 134–141, 2014.

S Bocklandt, W Lin, ME Sehl, FJ Sánchez, JS Sinsheimer, S Horvath, and E Vilain. Epigenetic predictor of age. *PloS ONE*, 6 (6): e14821, 2011.

W Branicki, F Liu, K van Duijn, J Draus-Barini, E Pośpiech, S Walsh, T Kupiec, A Wojas-Pelc, and M Kayser. Model-based prediction of human hair color using DNA variants. *Human Genetics*, 129 (4): 443–454, 2011.

CH Brenner. Difficulties in the estimation of ethnic affiliation. *American Journal of Human Genetics*, 62 (6): 1558, 1998.

C Brenner. Understanding y haplotype matching probability. *Forensic Science International: Genetics*, 8: 233–243, 2014.

CH Brenner. Forensic genetics: mathematics. *eLS*, 2: 513–519, 2006.

CH Brenner and BS Weir. Issues and strategies in the DNA identification of world trade center victims. *Theoretical Population Biology*, 63 (3): 173–178, 2003.

J-A Bright, JM Curran, and JS Buckleton. Relatedness calculations for linked loci incorporating subpopulation effects. *Forensic Science International: Genetics*, 7 (3): 380–383, 2013a.

J-A Bright, D Taylor, J Curran, and JS Buckleton. Degradation of forensic DNA profiles. *Australian Journal of Forensic Sciences*, 2013b, Published online, doi: 10.1080/00450618.2013.772235.

J-A Bright, D Taylor, JM Curran, and JS Buckleton. Developing allelic and stutter peak height models for a continuous method of DNA interpretation. *Forensic Science International: Genetics*, 7 (2): 296–304, 2013c.

J Buckleton, H Kelly, J-A Bright, D Taylor, T Tvedebrink, and JM Curran. Utilising allelic dropout probabilities estimated by logistic regression in casework. *Forensic Science International: Genetics*, 9: 9–11, 2014.

J Buckleton, C Triggs, and S Walsh. *Forensic DNA Evidence Interpretation*. CRC Press, 2005.

JM Butler. *Forensic DNA Typing: Biology, Technology, and Genetics of STR Markers*, 2nd ed. Academic Press, 2005.

JM Butler. *Fundamentals of Forensic DNA Typing*. Elsevier Academic Press, 2010.

J Butler. *Advanced Topics in Forensic DNA Typing: Interpretation*. Academic Press, 2014.

JM Butler, E Buel, F Crivellente, and BR McCord. Forensic DNA typing by capillary electrophoresis using the ABI prism 310 and 3100 genetic analyzers for STR analysis. *Electrophoresis*, 25 (10-11): 1397–1412, 2004.

B Caddy, G Taylor, and A Linacre. *A Review of the Science of Low Template DNA Analysis*. UK Home Office, 2008.

P Calabrese and R Durrett. Dinucleotide repeats in the drosophila and human genomes have complex, length-dependent mutation processes. *Molecular Biology and Evolution*, 20 (5): 715–725, 2003.

D Cavallini and F Corradi. Forensic identification of relatives of individuals included in a database of DNA profiles. *Biometrika*, 93 (3): 525–536, 2006.

R Chakraborty, MR Srinivasan, and SP Daiger. Evaluation of standard error and confidence interval of estimated multilocus genotype probabilities, and their implications in DNA forensics. *American Journal of Human Genetics*, 52 (1): 60, 1993.

A Chakravarti and CC Li. Estimating the prior probability of paternity from the results of exclusion tests. *Forensic Science International*, 24 (2): 143–147, 1984.

C Champod. Dna transfer: informed judgment or mere guesswork? *Frontiers in Genetics*, 4: 300, 2013.

BC Christensen, EA Houseman, CJ Marsit, S Zheng, MR Wrensch, JL Wiemels, HH Nelson, MR Karagas, JF Padbury, R Bueno, et al. Aging and environmental exposures alter tissue-specific DNA methylation dependent upon CPG island context. *PLoS Genetics*, 5 (8): e1000602, 2009.

P Claes, DK Liberton, K Daniels, KM Rosana, EE Quillen, LN Pearson, B McEvoy, M Bauchet, AA Zaidi, W Yao, et al. Modeling 3d facial shape from DNA. *PLoS Genetics*, 10 (3): e1004224, 2014.

TM Clayton, JP Whitaker, R Sparkes, and P Gill. Analysis and interpretation of mixed forensic stains using DNA STR profiling. *Forensic Science International*, 91 (1): 55–70, 1998.

R Cook, IW Evett, G Jackson, PJ Jones, and JA Lambert. A hierarchy of propositions: deciding which level to address in casework. *Science & Justice*, 38 (4): 231–239, 1998.

R Cowell, T Graversen, S Lauritzen, and J Mortera. Analysis of DNA mixtures with artefacts. *Journal of the Royal Statistical Society: Series A (Statistics in Society)*, 64 (1): 1–48, 2015. DOI: 10.1111/rssc.12071.

RG Cowell, T Graversen, SL Lauritzen, and J Mortera. Analysis of forensic DNA mixtures with artefacts. *Analysis*, 5: 6, 2014b.

R Cowell, S Lauritzen, and J Mortera. A gamma model for DNA mixture analyses. *Bayesian Analysis*, 2 (2): 333–348, 2007.

JM Curran, JS Buckleton, CM Triggs, and BS Weir. Assessing uncertainty in DNA evidence caused by sampling effects. *Science & Justice*, 42 (1): 29–37, 2002.

J Curran, P Gill, and M Bill. Interpretation of repeat measurement DNA evidence allowing for multiple contributors and population substructure. *Forensic Science International*, 148 (1): 47–53, 2005.

JM Curran, CM Triggs, J Buckleton, and BS Weir. Interpreting DNA mixtures in structured populations. *Journal of Forensic Sciences*, 44: 987–995, 1999.

AP Dawid. Comment on stockmarr's 'likelihood ratios for evaluating DNA evidence when the suspect is found through a database search' (with response by A. Stockmarr). *Biometrics*, 57: 976–980, 2001.

AP Dawid. Which likelihood ratio?(comment on 'why the effect of prior odds should accompany the likelihood ratio when reporting DNA evidence', by ronald meester and marjan sjerps). *Law, Probability and Risk*, 3 (1): 65–71, 2004.

AP Dawid and J Mortera. Coherent analysis of forensic identification evidence. *Journal of the Royal Statistical Society: Series B (Methodological)*, 58: 425–443, 1996.

AP Dawid and J Mortera. Forensic identification with imperfect evidence. *Biometrika*, 85 (4): 835–849, 1998.

AP Dawid, J Mortera, and VL Pascali. Non-fatherhood or mutation? A probabilistic approach to parental exclusion in paternity testing. *Forensic Science International*, 124 (1): 55–61, 2001.

AP Dawid, J Mortera, VL Pascali, and D Van Boxel. Probabilistic expert systems for forensic inference from genetic markers. *Scandinavian Journal of Statistics*, 29 (4): 577–595, 2002.

AP Dawid, J Mortera, and P Vicard. Object-oriented bayesian networks for complex forensic DNA profiling problems. *Forensic Science International*, 169 (2): 195–205, 2007.

P Donnelly and RD Friedman. DNA database searches and the legal consumption of scientific evidence. *Michigan Law Review*, 97 (4):931–984, 1999.

G Dørum, Ø Bleka, P Gill, H Haned, L Snipen, S Sæbø, and T Egeland. Exact computation of the distribution of likelihood ratios with forensic applications. *Forensic Science International: Genetics*, 9: 93–101, 2014.

T Egeland, PF Mostad, B Mevåg, and M Stenersen. Beyond traditional paternity and identification cases: selecting the most probable pedigree. *Forensic Science International*, 110 (1): 47–59, 2000.

R Eggleston. *Evidence, Proof, and Probability*, 2nd ed. Weidenfeld &; Nicolson, 1983.

IW Evett. Avoiding the transposed conditional. *Science & Justice*, 35 (2): 127–131, 1995.

I Evett, C Buffery, G Wilcott, and D Stoney. A guide to interpreting single locus profiles of DNA mixtures in forensic cases. *Journal of the Forensic Science Society*, 31: 41–47, 1991.

I Evett, P Gill, and J Lambert. Taking account of peak areas when interpreting mixed DNA profiles. *Journal of Forensic Science*, 43: 62–69, 1998.

IW Evett, G Jackson, and JA Lambert. More on the hierarchy of propositions: exploring the distinction between explanations and propositions. *Science & Justice*, 40 (1): 3–10, 2000.

I Evett and S Pope. Science of mixed results. *Law Society Gazette*, 2013. Available online at: http://www.lawgazette.co.uk/5036961.article.

I Evett and B Weir. *Interpreting DNA Evidence: Statistical Genetics for Forensic Scientists*. Sinauer Associates: Sunderland, MA, 1998.

MO Finkelstein and WB Fairley. A bayesian approach to identification evidence. *Harvard Law Review*, 83: 489–517, 1970a.

MO Finkelstein and WB Fairley. Comment on trial by mathematics. *Harvard Law Review*, 84: 1801, 1970b.

WK Fung. User-friendly programs for easy calculations in paternity testing and kinship determinations. *Forensic Science International*, 136 (1): 22–34, 2003.

WK Fung, A Carracedo, and Y-Q Hu. Testing for kinship in a subdivided population. *Forensic Science International*, 135 (2): 105–109, 2003.

WK Fung and Y-Q Hu. Interpreting forensic DNA mixtures: allowing for uncertainty in population substructure and dependence. *Journal of the Royal Statistical Society: Series A (Statistics in Society)*, 163 (2): 241–254, 2000.

WK Fung and Y-Q Hu. The statistical evaluation of DNA mixtures with contributors from different ethnic groups. *International Journal of Legal Medicine*, 116 (2): 79–86, 2002.

WK Fung and Y-Q Hu. Interpreting DNA mixtures with related contributors in subdivided populations. *Scandinavian Journal of Statistics*, 31 (1): 115–130, 2004.

P Gill. Does an English appeal court ruling increase the risks of miscarriages of justice when complex DNA profiles are searched against the national DNA database? *Forensic Science International*, 13: 167–175, 2014a.

P Gill. *Misleading DNA evidence: reasons for miscarriages of justice*. Academic Press, 2014b.

P Gill, C Brenner, J Buckleton, A Carracedo, M Krawczak, W Mayr, N Morling, M Prinz, P Schneider, and B Weir. DNA Commission of the International Society of Forensic Genetics: recommendations on the interpretation of mixtures. *Forensic Science International*, 160 (2-3): 90–101, 2006. ISSN: 0379-0738.

P Gill and H Haned. A new methodological framework to interpret complex DNA profiles using likelihood ratios. *Forensic Science International: Genetics*, 7 (2): 251–263, 2013. ISSN: 1872-4973.

P Gill, PL Ivanov, C Kimpton, R Piercy, N Benson, G Tully, I Evett, E Hagelberg, and K Sullivan. Identification of the remains of the romanov family by DNA analysis. *Nature Genetics*, 6 (2): 130–135, 1994.

P Gill, A Kirkham, and J Curran. LoComatioN: a software tool for the analysis of low copy number DNA profiles. *Forensic Science International*, 166: 128–138, 2007.

P Gill, R Sparkes, R Pinchin, T Clayton, J Whitaker, and J Buckleton. Interpreting simple STR mixtures using allele peak areas. *Forensic Science International*, 91 (1): 41–53, 1998. ISSN: 0379-0738.

P Gill, DJ Werrett, B Budowle, and R Guerrieri. An assessment of whether SNPs will replace STRs in national DNA databases–joint considerations of the DNA working group of the European Network of Forensic Science Institutes (ENFSI) and the Scientific Working Group on DNA Analysis Methods (SWGDAM). *Science & Justice: Journal of the Forensic Science Society*, 44 (1): 51, 2004.

P Gill, J Whitaker, C Flaxman, N Brown, and J Buckleton. An investigation of the rigor of interpretation rules for STRs derived from less than 100 pg of DNA. *Forensic Science International*, 112: 17–40, 2000.

WG Giusti and T Adriano. Synthesis and characterization of 5′-fluorescent-dye-labeled oligonucleotides. *Genome Research*, 2 (3): 223–227, 1993.

DW Gjertson, CH Brenner, MP Baur, A Carracedo, F Guidet, JA Luque, R Lessig, WR Mayr, VL Pascali, M Prinz, et al. ISFG: recommendations on biostatistics in paternity testing. *Forensic Science International: Genetics*, 1 (3): 223–231, 2007.

Globalfiler. Life technologies. Online, October 2014.

IJ Good. *Weight of Evidence and the Bayesian Likelihood Ratio*. Ellis Horwood: Chichester, 1991.

IJ Good. When batterer turns murderer. *Nature*, 375 (6532): 541–541, 1995.

M Goray, JR Mitchell, and RAH van Oorschot. Evaluation of multiple transfer of DNA using mock case scenarios. *Legal Medicine*, 14 (1): 40–46, 2012.

K Grisedale and A van Daal. Comparison of STR profiling from low template DNA extracts with and without the consensus profiling method. *Investigative Genetics*, 3: 1–9, 2012.

I Halder, M Shriver, M Thomas, JR Fernandez, and T Frudakis. A panel of ancestry informative markers for estimating individual biogeographical ancestry and admixture from four continents: utility and applications. *Human Mutation*, 29 (5): 648–658, 2008.

H Haned, L Pène, JR Lobry, AB Dufour, and D Pontier. Estimating the number of contributors to forensic DNA mixtures: does maximum likelihood perform better than maximum allele count? *Journal of Forensic Sciences*, 56 (1): 23–28, 2011.

G Hellenthal, GBJ Busby, G Band, JF Wilson, C Capelli, D Falush, and S Myers. A genetic atlas of human admixture history. *Science*, 343 (6172): 747–751, 2014.

AB Hepler and BS Weir. Object-oriented bayesian networks for paternity cases with allelic dependencies. *Forensic Science International: Genetics*, 2 (3): 166–175, 2008.

AM Hillmer, FF Brockschmidt, S Hanneken, S Eigelshoven, M Steffens, A Flaquer, S Herms, T Becker, A-K Kortüm, DR Nyholt, et al. Susceptibility variants for male-pattern baldness on chromosome 20p11. *Nature Genetics*, 40 (11): 1279–1281, 2008.

Y-Q Hu and WK Fung. Evaluating forensic DNA mixtures with contributors of different structured ethnic origins: a computer software. *International Journal of Legal Medicine*, 117 (4): 248–249, 2003.

K Ikematsu, R Tsuda, and I Nakasono. Gene response of mouse skin to pressure injury in the neck region. *Legal Medicine*, 8 (2): 128–131, 2006.

J Jia, Y-L Wei, C-J Qin, L Hu, L-H Wan, and C-X Li. Developing a novel panel of genome-wide ancestry informative markers for bio-geographical ancestry estimates. *Forensic Science International: Genetics*, 8 (1): 187–194, 2014.

MA Jobling, M Hurles, and C Tyler-Smith. *Human Evolutionary Genetics: Origins, Peoples & Disease*. Garland Science, 2004.

MA Jobling, A Pandya, and C Tyler-Smith. The y chromosome in forensic analysis and paternity testing. *International Journal of Legal Medicine*, 110 (3): 118–124, 1997.

J Juusola and J Ballantyne. MRNA profiling for body fluid identification by multiplex quantitative RT-PCR*. *Journal of Forensic Sciences*, 52 (6): 1252–1262, 2007.

N Kaur, AE Fonneløp, and T Egeland. Regression models for DNA-mixtures. *Forensic Science International: Genetics*, 11: 105–110, 2014.

DH Kaye. The probability of an ultimate issue: the strange cases of paternity testing. *Iowa Law Review*, 75: 75, 1989.

DH Kaye. Questioning a courtroom proof of the uniqueness of fingerprints. *International Statistical Review*, 71 (3): 521–533, 2003.

M Kayser and P de Knijff. Improving human forensics through advances in genetics, genomics and molecular biology. *Nature Reviews Genetics*, 12 (3): 179–192, 2011.

H Kelly, J-A Bright, JS Buckleton, and JM Curran. Identifying and modelling the drivers of stutter in forensic DNA profiles. *Australian Journal of Forensic Sciences*, 46 (2): 194–203, 2013. DOI: 10.1080/00450618.2013.808697.

M Kimura. Preponderance of synonymous changes as evidence for the neutral theory of molecular evolution. *Nature*, 267: 275–276, 1977.

CM Koch and W Wagner. Epigenetic-aging-signature to determine age in different tissues. *Aging (Albany NY)*, 3 (10): 1018, 2011.

JJ Koehler. DNA matches and statistics: important questions, surprising answers. *Judicature*, 76: 222, 1992.

JJ Koehler. Error and exaggeration in the presentation of DNA evidence at trial. *Jurimetrics Journal*, 34: 21, 1993.

JJ Koehler. On conveying the probative value of DNA evidence: frequencies, likelihood ratios, and error rates. *University of Colorado Law Review*, 67: 859, 1996.

JJ Koehler. When are people persuaded by DNA match statistics? *Law and Human Behavior*, 25 (5): 493, 2001.

K Krjutškov, T Viltrop, P Palta, E Metspalu, E Tamm, S Suvi, K Sak, A Merilo, H Sork, R Teek, et al. Evaluation of the 124-plex SNP typing microarray for forensic testing. *Forensic Science International: Genetics*, 4 (1): 43–48, 2009.

Y Lai and F Sun. The relationship between microsatellite slippage mutation rate and the number of repeat units. *Molecular Biology and Evolution*, 20 (12): 2123–2131, 2003.

JA Lambert and IW Evett. The impact of recent judgements on the presentation of DNA evidence. *Science & Justice*, 38 (4): 266–270, 1998.

SL Lauritzen and NA Sheehan. Graphical models for genetic analyses. *Statistical Science*, 18 (4): 489–514, 2003.

CC Li and A Chakravarti. An expository review of two methods of calculating the paternity probability. *American Journal of Human Genetics*, 43 (2): 197, 1988.

J Liu, JM Lewohl, PR Dodd, PK Randall, RA Harris, and RD Mayfield. Gene expression profiling of individual cases reveals consistent transcriptional changes in alcoholic human brain. *Journal of Neurochemistry*, 90 (5): 1050–1058, 2004.

F Liu, F van der Lijn, C Schurmann, G Zhu, MM Chakravarty, PG Hysi, A Wollstein, O Lao, M de Bruijne, MA Ikram, et al. A genome-wide association study identifies five loci influencing facial morphology in Europeans. *PLoS Genetics*, 8 (9): e1002932, 2012.

K Lohmueller and N Rudin. Calculating the weight of evidence in low-template Forensic DNA casework. *Journal of Forensic Science*, 12017: 1–7, 2012. ISSN: 1556-4029.

A Lowe, C Murray, J Whitaker, G Tully, and P Gill. The propensity of individuals to deposit DNA and secondary transfer of low level DNA from individuals to inert surfaces. *Forensic Science International*, 129 (1): 25–34, 2002.

CN Maguire, LA McCallum, C Storey, and JP Whitaker. Familial searching: a specialist forensic DNA profiling service utilising the national DNA database® to identify unknown offenders via their relatives – the UK experience. *Forensic Science International: Genetics*, 8 (1): 1–9, 2014.

PJ Maiste and BS Weir. A comparison of tests for independence in the FBI RFLP data bases. In *Human Identification: The Use of DNA Markers*, pages 125–138. Springer, 1995.

J Marchini, LR Cardon, MS Phillips, and P Donnelly. The effects of human population structure on large genetic association studies. *Nature Genetics*, 36 (5): 512–517, 2004.

K Martire, K R, I Watkins, M Sayle, and B Newell. The expression and interpretation of uncertain forensic science evidence: verbal equivalence, evidence strength and the weak evidence effect. *Law and Human Behavior*, 37: 197–207, 2013.

G Meakin and A Jamieson. DNA transfer: review and implications for casework. *Forensic Science International: Genetics*, 7 (4): 434–443, 2013.

R Meester and M Sjerps. The evidential value in the DNA database search controversy and the two-stain problem. *Biometrics*, 59 (3): 727–732, 2003.

R Meester and M Sjerps. Why the effect of prior odds should accompany the likelihood ratio when reporting DNA evidence. *Law, Probability and Risk*, 3 (1): 51–62, 2004.

A Mitchell, J Tamariz, K Connell, N Ducasse, Z Budimlija, M Prinz, and T Caragine. Validation of a DNA mixture statistics tool incorporating allelic drop-out and drop-in. *Forensic Science International: Genetics*, 6: 749–761, 2012.

GS Morrison. The likelihood-ratio framework and forensic evidence in court: a response to R v T. *International Journal of Evidence & Proof*, 16 (1): 1–29, 2012.

J Mortera, AP Dawid, and SL Lauritzen. Probabilistic expert systems for DNA mixture profiling. *Theoretical Population Biology*, 63 (3): 191–205, 2003.

S Myles, M Somel, K Tang, J Kelso, and M Stoneking. Identifying genes underlying skin pigmentation differences among human populations. *Human Genetics*, 120 (5): 613–621, 2007.

National Research Council. *The Evaluation of Forensic DNA Evidence*. National Academies Press: Washington, DC, 1996.

National Research Council. *Strengthening Forensic Science in the United States: a Path Forward*. National Academies Press: Washington, DC, 2009.

National Research Council, United States of America. DNA technology in forensic science, 1992.

C Neumann, I Evett, and J Skerrett. Quantifying the weight of evidence from a forensic fingerprint comparison: a new paradigm. *Journal of the Royal Statistical Society: Series A (Statistics in Society)*, 175 (2): 371–415, 2012. ISSN: 09641998.

J Newell-Price, AJL Clark, and P King. Dna methylation and silencing of gene expression. *Trends in Endocrinology and Metabolism*, 11 (4): 142–148, 2000.

Office of the US Inspector General. *A Review of the FBIs Handling of the Brandon Mayfield Case: Special Report*. US Department of Justice: Washington, DC, 2006.

W Parson, A Brandstätter, A Alonso, N Brandt, B Brinkmann, A Carracedo, D Corach, O Froment, I Furac, T Grzybowski, et al. The EDNAP mitochondrial DNA population database (EMPOP) collaborative exercises: organisation, results and perspectives. *Forensic Science International*, 139 (2): 215–226, 2004.

J Perez, AA Mitchell, N Ducasse, J Tamariz, and T Caragine. Estimating the number of contributors to two-, three-, and four-person mixtures containing DNA in high template and low template amounts. *Croatian Medical Journal*, 52 (3): 314–326, 2011.

M Perlin, M Legler, C Spencer, J Smith, W Allan, J Belrose, and B Duceman. Validating TrueAllele (R) DNA mixture interpretation. *Journal of Forensic Science*, 56 (6): 1430–1447, 2011a. ISSN: 0022-1198.

MW Perlin, MM Legler, CE Spencer, JL Smith, WP Allan, JL Belrose, and BW Duceman. Validating trueallele® DNA mixture interpretation*,. *Journal of Forensic Sciences*, 56 (6): 1430–1447, 2011b.

M Perlin and A Sinelnikov. An information gap in DNA evidence interpretation. *PloS ONE*, 4 (12): e8327, 2009. ISSN: 1932-6203.

M Perlin and B Szabady. Linear mixture analysis: a mathematical approach to resolving mixed DNA samples. *Journal of Forensic Science*, 46: 1372–1378, 2001.

CM Pfeifer, R Klein-Unseld, M Klintschar, and P Wiegand. Comparison of different interpretation strategies for low template DNA mixtures. *Forensic Science International: Genetics*, 6 (6): 716–722, 2012.

C Phillips, D Ballard, P Gill, DS Court, Á Carracedo, and MV Lareu. The recombination landscape around forensic strs: accurate measurement of genetic distances between syntenic STR pairs using hapmap high density SNP data. *Forensic Science International: Genetics*, 6 (3): 354–365, 2012.

C Phillips, M Gelabert-Besada, L Fernandez-Formoso, M García-Magari nos, C Santos, M Fondevila, D Ballard, DS Court, Á Carracedo, and MV Lareu. New turns from old stars': enhancing the capabilities of forensic short tandem repeat analysis. *Electrophoresis*, 35 (2123): 2014.

C Phillips, A Salas, JJ Sanchez, M Fondevila, A Gomez-Tato, J Alvarez-Dios, M Calaza, M Casares de Cal, D Ballard, MV Lareu, et al. Inferring ancestral origin using a single multiplex assay of ancestry-informative marker SNPs. *Forensic Science International: Genetics*, 1 (3): 273–280, 2007.

NJ Port, VL Bowyer, EAM Graham, MS Batuwangala, and GN Rutty. How long does it take a static speaking individual to contaminate the immediate environment? *Forensic Science, Medicine and Pathology*, 2 (3): 157–163, 2006.

Promega. Powerplex® fusion system. Online, June 2014.

R Puch-Solis, A Kirkham, P Gill, J Read, S Watson, and D Drew. Practical determination of the low template DNA threshold. *Forensic Science International: Genetics*, 5 (5): 422–7, 2011. ISSN: 1878-0326.

R Puch-Solis, L Rodgers, A Mazumder, I Evett, S Pope, J Curran, and DJ Balding. Evaluating forensic DNA profiles using peak heights, allowing for multiple donors, allelic dropout and stutters. *Forensic Science International: Genetics*, 7 (1): 52–54, 2013a.

R Puch-Solis, L Rodgers, A Mazumder, S Pope, I Evett, J Curran, and D Balding. Evaluating forensic DNA profiles using peak heights, allowing for multiple donors, allelic dropout and stutters. *Forensic Science International: Genetics*, 7 (5): 555–563, 2013b.

J Pueschel. The application of bayesian hierarchical models to DNA profiling data. PhD Thesis, University College of London, 2001.

B Rannala and JL Mountain. Detecting immigration by using multilocus genotypes. *Proceedings of the National Academy of Sciences of the United States of America*, 94 (17): 9197–9201, 1997.

M Redmayne, P Roberts, C Aitken, and G Jackson. Forensic science evidence in question. *Criminal Law Review*. 5: 347–356, 2011.

A Rennison. *Report Into the Circumstances of a Complaint Received from the Greater Manchester Police on 7 March 2012 Regarding DNA Evidence Provided by LGC Forensics*. UK Home Office, 2012.

B Robertson and T Vignaux. *Interpreting Evidence: Evaluating Forensic Science in the Court-room.* John Wiley & Sons, Ltd, 1995.

B Robertson, GA Vignaux, and CEH Berger. Extending the confusion about bayes. *Modern Law Review*, 74 (3): 444–455, 2011.

K Roeder et al. DNA fingerprinting: a review of the controversy. *Statistical Science*, 9 (2): 222–247, 1994.

RV Rohlfs, E Murphy, YS Song, and Slatkin M. The influence of relatives on the efficiency and error rate of familial searching. *PLoS ONE*, 8 (8): e70495, 2013.

C Romualdi, D Balding, IS Nasidze, G Risch, M Robichaux, ST Sherry, M Stoneking, MA Batzer, and G Barbujani. Patterns of human diversity, within and among continents, inferred from biallelic DNA polymorphisms. *Genome Research*, 12 (4): 602–612, 2002.

DJ Rose Jr and JW Jorgenson. Characterization and automation of sample introduction methods for capillary zone electrophoresis. *Analytical Chemistry*, 60 (7): 642–648, 1988.

NA Rosenberg, JK Pritchard, JL Weber, HM Cann, KK Kidd, LA Zhivotovsky, and MW Feldman. Genetic structure of human populations. *Science*, 298 (5602): 2381–2385, 2002.

N Rudin and K Inman. *An Introduction to Forensic DNA Analysis*, 2nd ed., volume 3. CRC Press, 2001.

N Rudin and K Inman. *The Experience Fallacy*. California Association of Criminalists News, 4th Quarter. pages 10–13, 2010. Available online at http://www.cacnews.org/news/4thq10.pdf.

JJ Sanchez, C Phillips, C Børsting, K Balogh, M Bogus, M Fondevila, CD Harrison, E Musgrave-Brown, A Salas, D Syndercombe-Court, et al. A multiplex assay with 52 single nucleotide polymorphisms for human identification. *Electrophoresis*, 27 (9): 1713–1724, 2006.

DJ Schaid. Linkage disequilibrium testing when linkage phase is unknown. *Genetics*, 166 (1): 505–512, 2004.

PM Schneider, R Fimmers, W Keil, G Molsberger, D Patzelt, W Pflug, T Rothämel, H Schmitter, H Schneider, and B Brinkmann. The German Stain Commission: recommendations for the interpretation of mixed stains. *International Journal of Legal Medicine*, 123 (1): 1–5, 2009.

M Setzer, J Juusola, and J Ballantyne. Recovery and stability of RNA in vaginal swabs and blood, semen, and saliva stains. *Journal of Forensic Sciences*, 53 (2): 296–305, 2008.

MD Shriver, MW Smith, and L Jin. Reply to brenner. *American Journal of Human Genetics*, 62 (6): 1560–1561, 1998.

MD Shriver, MW Smith, L Jin, A Marcini, JM Akey, R Deka, and RE Ferrell. Ethnic-affiliation estimation by use of population-specific DNA markers. *American Journal of Human Genetics*, 60 (4): 957, 1997.

MJ Sjerps and CEH Berger. How clear is transparent? Reporting expert reasoning in legal cases. *Law, Probability and Risk*, 11 (4): 317–329, 2012.

CD Steele and DJ Balding. Statistical evaluation of forensic DNA profile evidence. *Annual Review of Statistics and Its Application*, 1: 20–21, 2014a.

CD Steele and DJ Balding. Choice of population database for forensic DNA profile analysis. *Science & Justice*, 54: 487–493, 2014b.

CD Steele, DS Court, and DJ Balding. Worldwide FST estimates relative to five continental-scale populations. *Annals of Human Genetics*, 78 (6): 468–477, 2014a.

CD Steele, M Greenhalgh, and DJ Balding. Verifying likelihoods for low template DNA profiles using multiple replicates. *Forensic Science International: Genetics*, 13: 82–89, 2014b.

A Stockmarr. Likelihood ratios for evaluating DNA evidence when the suspect is found through a database search. *Biometrics*, 55 (3): 671–677, 1999.

R Straussman, D Nejman, D Roberts, I Steinfeld, B Blum, N Benvenisty, I Simon, Z Yakhini, and H Cedar. Developmental programming of CPG Island methylation profiles in the human genome. *Nature Structural & Molecular Biology*, 16 (5): 564–571, 2009.

S Subramanian, RK Mishra, and L Singh. Genome-wide analysis of BKM sequences (gata repeats): predominant association with sex chromosomes and potential role in higher order chromatin organization and function. *Bioinformatics*, 19 (6): 681–685, 2003.

R Szibor, M Krawczak, S Hering, J Edelmann, E Kuhlisch, and D Krause. Use of X-linked markers for forensic purposes. *International Journal of Legal Medicine*, 117 (2): 67–74, 2003.

M Szyf. DNA methylation, the early-life social environment and behavioral disorders. *Journal of Neurodevelopmental Disorders*, 3 (3): 238–249, 2011.

F Taroni, A Biedermann, S Bozza, P Garbolino, and C Aitken. *Bayesian Networks for Probabilistic Inference and Decision Analysis in Forensic Science*. John Wiley & Sons, 2014.

D Taylor. Using continuous DNA interpretation methods to revisit likelihood ratio behaviour. *Forensic Science International: Genetics*, 11: 144–153, 2014.

D Taylor, J-A Bright, and J Buckleton. The interpretation of single source and mixed DNA profiles. *Forensic Science International: Genetics*, 7 (5): 516–528, 2013. ISSN: 18724973.

WC Thompson. Are juries competent to evaluate statistical evidence? *Law and Contemporary Problems*, 52: 9–41, 1989.

WC Thompson and EL Schumann. Interpretation of statistical evidence in criminal trials: the prosecutor's fallacy and the defense attorney's fallacy. *Law and Human Behavior*, 11 (3): 167, 1987.

WC Thompson, F Taroni, and CGG Aitken. How the probability of a false positive affects the value of DNA evidence. *Journal of Forensic Sciences*, 48 (1): 47–54, 2003.

Y Torres, I Flores, V Prieto, M López-Soto, MJ Farfán, A Carracedo, and P Sanz. DNA mixtures in forensic casework: a 4-year retrospective study. *Forensic Science International*, 134 (2): 180–186, 2003.

LH Tribe. Trial by mathematics: precision and ritual in the legal process. *Harvard Law Review*, 84: 1329, 1970.

G Tully, W Bär, B Brinkmann, A Carracedo, P Gill, N Morling, W Parson, and P Schneider. Considerations by the European DNA profiling (EDNAP) group on the working practices, nomenclature and interpretation of mitochondrial DNA profiles. *Forensic Science International*, 124 (1): 83–91, 2001.

T Tvedebrink, P Eriksen, H Mogensen, and N Morling. Estimating the probability of allelic drop-out of STR alleles in forensic genetics. *Forensic Science International: Genetics*, 3: 222–226, 2009.

T Tvedebrink, P Eriksen, H Mogensen, and N Morling. Evaluating the weight of evidence by using quantitative short tandem repeat data in DNA mixtures. *Applied Statistics*, 89: 855–874, 2010.

T Tvedebrink, P Eriksen, H Mogensen, and N Morling. Statistical model for degraded DNA samples and adjusted probabilities for allelic drop-out. *Forensic Science International: Genetics*, 6: 97–101, 2012.

UK Association of Forensic Science Providers. Standards for the formulation of evaluative forensic science expert opinion. *Science & Justice*, 49: 161–164, 2009.

UK Parliament. Protection of Freedoms Act 2012, 2013.

B Ulery, R Hicklin, J Buscaglia, and M Roberts. Accuracy and reliability of forensic latent fingerprint decisions. *Proceedings of the National academy of Sciences of the United States of America*, 108: 7733–38, 2010.

M Vennemann and A Koppelkamm. MRNA profiling in forensic genetics I: Possibilities and limitations. *Forensic Science International*, 203 (1): 71–75, 2010.

A Vidaki, B Daniel, and Denise Syndercombe Court. Forensic DNA methylation profiling – potential opportunities and challenges. *Forensic Science International: Genetics*, 7 (5): 499–507, 2013.

S Walsh, F Liu, KN Ballantyne, M van Oven, O Lao, and M Kayser. Irisplex: a sensitive DNA tool for accurate prediction of blue and brown eye colour in the absence of ancestry information. *Forensic Science International: Genetics*, 5 (3): 170–180, 2011.

S Walsh, F Liu, A Wollstein, L Kovatsi, A Ralf, A Kosiniak-Kamysz, W Branicki, and M Kayser. The hirisplex system for simultaneous prediction of hair and eye colour from DNA. *Forensic Science International: Genetics*, 7 (1): 98–115, 2013.

BS Weir. The effects of inbreeding on forensic calculations. *Annual Review of Genetics*, 28 (1): 597–621, 1994.

BS Weir. Forensics. In DJ Balding, M Bishop, and C Cannings, editors, *Handbook of Statistical Genetics*, volume 2, 3rd ed., chapter 43, pages 1368–92. John Wiley & Sons, Ltd, 2007.

BS Weir and WG Hill. Estimating F-statistics. *Annual Review of Genetics*, 36 (1): 721–750, 2002.

BS Weir and J Ott. Genetic data analysis II. *Trends in Genetics*, 13 (9): 379, 1997.

AA Westen, AS Matai, JFJ Laros, HC Meiland, M Jasper, WJF de Leeuw, P de Knijff, and T Sijen. Tri-allelic SNP markers enable analysis of mixed and degraded DNA samples. *Forensic Science International: Genetics*, 3 (4): 233–241, 2009.

JC Whittaker, RM Harbord, N Boxall, I Mackay, G Dawson, and RM Sibly. Likelihood-based estimation of microsatellite mutation rates. *Genetics*, 164 (2): 781–787, 2003.

L Wild and JM Flanagan. Genome-wide hypomethylation in cancer may be a passive consequence of transformation. *Biochimica et Biophysica Acta (BBA)-Reviews on Cancer*, 1806 (1): 50–57, 2010.

IJ Wilson, ME Weale, and DJ Balding. Inferences from DNA data: population histories, evolutionary processes and forensic match probabilities. *Journal of the Royal Statistical Society: Series A (Statistics in Society)*, 166 (2): 155–188, 2003.

S Wright. The genetic structure of populations. *Annals of Eugenics*, 15: 313–354, 1951.

W Wu, H Hao, Q Liu, X Han, Y Wu, J Cheng, and D Lu. Analysis of linkage and linkage disequilibrium for syntenic STRs on 12 chromosomes. *International Journal of Legal Medicine*, 128 (5): 735-739, 2014.

X Xu, M Peng, Z Fang, and X Xu. The direction of microsatellite mutations is dependent upon allele length. *Nature Genetics*, 24 (4): 396–399, 2000.

N Yang, H Li, LA Criswell, PK Gregersen, ME Alarcon-Riquelme, R Kittles, R Shigeta, G Silva, PI Patel, JW Belmont, et al. Examination of ancestry and ethnic affiliation using highly informative diallelic DNA markers: application to diverse and admixed populations and implications for clinical epidemiology and forensic medicine. *Human Genetics*, 118 (3-4): 382–392, 2005.

MT Zarrabeitia, JA Riancho, MV Lareu, F Leyva-Cobián, and A Carracedo. Significance of micro-geographical population structure in forensic cases: a bayesian exploration. *International Journal of Legal Medicine*, 117 (5): 302–305, 2003.

D Zaykin, L Zhivotovsky, and BS Weir. Exact tests for association between alleles at arbitrary numbers of loci. *Genetica*, 96 (1-2): 169–178, 1995.

T Zerjal, Y Xue, G Bertorelle, RS Wells, W Bao, S Zhu, R Qamar, Q Ayub, A Mohyuddin, S Fu, et al. The genetic legacy of the mongols. *American Journal of Human Genetics*, 72 (3): 717–721, 2003.

Index

Weight-of-Evidence for Forensic DNA Profiles, Second Edition.
David J. Balding and Christopher D. Steele.
© 2015 John Wiley & Sons, Ltd. Published 2015 by John Wiley & Sons, Ltd.
Companion Website: www.wiley.com/go/balding/weight_of_evidence

Statistics in Practice

Human and Biological Sciences

Berger – Selection Bias and Covariate Imbalances in Randomized Clinical Trials

Berger and Wong – An Introduction to Optimal Designs for Social and Biomedical Research

Brown, Gregory, Twelves and Brown – A Practical Guide to Designing Phase II Trials in Oncology

Brown and Prescott – Applied Mixed Models in Medicine, Third Edition

Campbell and Walters – How to Design, Analyse and Report Cluster Randomised Trials in Medicine and Health Related Research

Carpenter and Kenward – Multiple Imputation and its Application

Carstensen – Comparing Clinical Measurement Methods

Chevret (Ed.) – Statistical Methods for Dose-Finding Experiments

Cooke – Uncertainty Modeling in Dose Response: Bench Testing Environmental Toxicity

Eldridge – A Practical Guide to Cluster Randomised Trials in Health Services Research

Ellenberg, Fleming and DeMets – Data Monitoring Committees in Clinical Trials: A Practical Perspective

Gould (Ed) – Statistical Methods for Evaluating Safety in Medical Product Development

Hauschke, Steinijans and Pigeot – Bioequivalence Studies in Drug Development: Methods and Applications

Källén – Understanding Biostatistics

Lawson, Browne and Vidal Rodeiro – Disease Mapping with Win-BUGS and MLwiN

Lesaffre, Feine, Leroux and Declerck – Statistical and Methodological Aspects of Oral Health Research

Lesaffre and Lawson – Bayesian Biostatistics

Lui – Binary Data Analysis of Randomized Clinical Trials with Noncompliance

Lui – Statistical Estimation of Epidemiological Risk

Marubini and Valsecchi – Analysing Survival Data from Clinical Trials and Observation Studies

Millar – Maximum Likelihood Estimation and Inference: With Examples in R, SAS and ADMB

Molenberghs and Kenward – Missing Data in Clinical Studies

Morton, Mengersen, Playford and Whitby – Statistical Methods for Hospital Monitoring with R

O'Hagan, Buck, Daneshkhah, Eiser, Garthwaite, Jenkinson, Oakley and Rakow – Uncertain Judgements: Eliciting Expert's Probabilities

O'Kelly and Ratitch – Clinical Trials with Missing Data: A Guide for Practitioners

Parmigiani – Modeling in Medical Decision Making: A Bayesian Approach

Pawlowsky-Glahn, Egozcue and Tolosana-Delgado – Modeling and Analysis of Compositional Data

Pintilie – Competing Risks: A Practical Perspective

Senn – Cross-over Trials in Clinical Research, Second Edition

Senn – Statistical Issues in Drug Development, Second Edition

Spiegelhalter, Abrams and Myles – Bayesian Approaches to Clinical Trials and Health-Care Evaluation

Walters – Quality of Life Outcomes in Clinical Trials and Health-Care Evaluation

Welton, Sutton, Cooper and Ades – Evidence Synthesis for Decision Making in Healthcare

Whitehead – Design and Analysis of Sequential Clinical Trials, Revised Second Edition

Whitehead – Meta-Analysis of Controlled Clinical Trials

Willan and Briggs – Statistical Analysis of Cost Effectiveness Data

Winkel and Zhang – Statistical Development of Quality in Medicine

Zhou, Zhou, Lui and Ding – Applied Missing Data Analysis in the Health Sciences

Earth and Environmental Sciences

Buck, Cavanagh and Litton – Bayesian Approach to Interpreting Archaeological Data

Chandler and Scott – Statistical Methods for Trend Detection and Analysis in the Environmental Statistics

Christie, Cliffe, Dawid and Senn (Eds.) – Simplicity, Complexity and Modelling

Gibbons, Bhaumik and Aryal – Statistical Methods for Groundwater Monitoring, 2nd Edition

Haas – Improving Natural Resource Management: Ecological and Political Models

Haas – Introduction to Probability and Statistics for Ecosystem Managers

Helsel – Nondetects and Data Analysis: Statistics for Censored Environmental Data

Illian, Penttinen, Stoyan and Stoyan – Statistical Analysis and Modelling of Spatial Point Patterns

Mateu and Muller (Eds) – Spatio-Temporal Design: Advances in Efficient Data Acquisition

McBride – Using Statistical Methods for Water Quality Management

Ofungwu – Statistical Applications for Environmental Analysis and Risk Assessment

Okabe and Sugihara – Spatial Analysis Along Networks: Statistical and Computational Methods

Pawlowsky-Glahn, Egozcue and Tolosana-Delgado – Modeling and Analysis of Compositional Data

Webster and Oliver – Geostatistics for Environmental Scientists, Second Edition

Wymer (Ed.) – Statistical Framework for RecreationalWater Quality Criteria and Monitoring

Industry, Commerce and Finance

Aitken – Statistics and the Evaluation of Evidence for Forensic Scientists, Second Edition

Balding and Steele – Weight-of-evidence for Forensic DNA Profiles, Second Edition

Brandimarte – Numerical Methods in Finance and Economics: A MATLAB-Based Introduction, Second Edition

Brandimarte and Zotteri – Introduction to Distribution Logistics

Chan – Simulation Techniques in Financial Risk Management

Coleman, Greenfield, Stewardson and Montgomery (Eds) – Statistical Practice in Business and Industry

Frisen (Ed.) – Financial Surveillance

Fung and Hu – Statistical DNA Forensics

Gusti Ngurah Agung – Time Series Data Analysis Using EViews

Jank and Shmueli – Modeling Online Auctions

Jank and Shmueli (Ed.) – Statistical Methods in e-Commerce Research

Lloyd – Data Driven Business Decisions

Kenett (Ed.) – Operational Risk Management: A Practical Approach to Intelligent Data Analysis

Kenett (Ed.) – Modern Analysis of Customer Surveys: With Applications using R

Kenett and Zacks – Modern Industrial Statistics: With Applications in R, MINITAB and JMP, Second Edition

Kruger and Xie – Statistical Monitoring of Complex Multivariate Processes: With Applications in Industrial Process Control

Lehtonen and Pahkinen – Practical Methods for Design and Analysis of Complex Surveys, Second Edition

Mallick, Gold, and Baladandayuthapani – Bayesian Analysis of Gene Expression Data

Ohser and Mücklich – Statistical Analysis of Microstructures in Materials Science

Pasiouras (Ed.) – Efficiency and Productivity Growth: Modelling in the Financial Services Industry

Pawlowsky-Glahn, Egozcue and Tolosana-Delgado – Modeling and Analysis of Compositional Data

Pfaff – Financial Risk Modelling and Portfolio Optimization with R

Pourret, Naim and Marcot (Eds) – Bayesian Networks: A Practical Guide to Applications

Rausand – Risk Assessment: Theory, Methods, and Applications

Ruggeri, Kenett and Faltin – Encyclopedia of Statistics and Reliability

Taroni, Biedermann, Bozza, Garbolino and Aitken – Bayesian Networks for Probabilistic Inference and Decision Analysis in Forensic Science, Second Edition

Taroni, Bozza, Biedermann, Garbolino and Aitken – Data Analysis in Forensic Science